Elementary Technical ENGLISH
Students' Book 2

Martin Webber and Jonathan Seath

Thomas Nelson and Sons Ltd
Nelson House   Mayfield Road
Walton-on-Thames   Surrey   KT12 5PL

51 York Place
Edinburgh   EH1 3JD

Thomas Nelson (Hong Kong) Ltd
Toppan Building 10/F
22A Westlands Road
Quarry Bay   Hong Kong

© Martin Webber and Jonathan Seath 1984

First published by Thomas Nelson and Sons Ltd
1984
ISBN 0-17-555354-8
NPN 10 9

All rights reserved. No part of this publication may be reproduced, copied or transmitted save with written permission or in accordance with the provisions of the Copyright, Design and Patents Act 1988, or under the terms of any licence permitting limited copying issued by the Copyright Licensing Agency, 90 Tottenham Court Road, London, W1P 9HE.

Any person who does any unauthorised act in relation to this publication may be liable to criminal prosecution and civil claims for damages.

Printed in Hong Kong

Some of the material in Unit 8 is reproduced by kind permission of the Engineering Industry Training Board.

Contents

Topics

1 Aircraft and Aircraft Engines, page 1

Operation and control

Word list, page 10

2 Levers and Machines, page 11

Force and effort
Basic arithmetic

Word list, page 20

3 Engineering Materials, page 21

Properties of engineering materials
Uses

Word list, page 32

4 Electricity, page 33

Simple circuits
Ohm's law
Maintenance
Generation

Word list, page 35

Language content

Some and most
Most piston engines in aircraft have four or eight cylinders.
Some have six or twelve.
Pronunciation of units of measurement
Direction of movement
Upwards, (anti-)clockwise, rotational

If – sentences expressing scientific truth
If water is heated to 100 °C, it boils.
Pronunciation of numbers and arithmetic operations
Four times two equals eight.
$4 \times 2 = 8$

Comparison
Copper is tougher than rubber.
Pronouns: *it* and *which*
Mild steel is less expensive than wrought iron because it is easier to produce.
Mild steel is a type of ferrous metal which contains 0.08 – 0.25% carbon.
Pronunciation of decimals
Reasons
Therefore and because

Location
In the middle, on the left
-ing form of verb with prepositions
After removing the plug leads . . .
Revision of simple present passive
Should
You should remove the plug lead.

5 Radio and Television, page 46

Waves and frequency
Microphones
Transmission of radio and television
    programmes

Connecting words
because, and, but, however
Pronunciation of units
Frequency is measured in cycles per second.
Nouns, verbs and adjectives
Vibration, vibrate, vibrating
Number and amount
A large number of electrons
A small amount of electricity

Word list, page 60

6 The History of Engines, page 61

Internal combustion
Diesel
Jet engines
Rockets
Inventions

Simple past – active and passive
Otto designed the internal combustion engine
    in 1870.
The blades were rotated.
Examples: *such as*
A fuel such as petrol.

Word list, page 73

7 Electrical Maintenance, page 74

Batteries

Structure
A battery consists of a number of cells.
Should
A battery should be kept clean.
Prepositions

Word list, page 84

8 Technical Books and Reports, page 85

Safety in the workshop
Motor vehicle maintenance
Tools
Welding
Electrical installations
Word list, page 97

Warnings
Don't use . . .
. . . must not be used . . .
Reporting faults
Why . . . ? Because . . .
The fuse needs replacing.

Comprehensive Word List, page 98

Unit 1
Aircraft and Aircraft Engines

Study Section 1.1

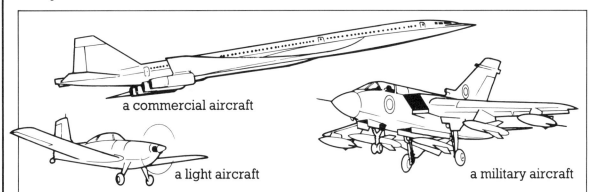

a commercial aircraft

a light aircraft

a military aircraft

Most military and commercial aircraft have jet engines. Most light aircraft have piston engines. Commercial aircraft generally have three or four engines. Light aircraft generally have one or two engines. Military aircraft have one, two, three or four engines.

Some helicopters have jet engines. Some have piston engines. There are commercial helicopters and there are military helicopters.

Engines produce power by burning fuel and air. Piston engines burn petrol. Jet engines burn kerosene. Piston engines have pistons and cylinders. Most piston engines in aircraft have four or eight cylinders but some have six or twelve. Jet engines do not have cylinders or pistons. They have turbines and compressors.

Practice 1

Are these statements true or false?
Correct the false ones.

For example:
Most light aircraft have jet engines.
No, most light aircraft have piston engines.

1. Most commercial aircraft have jet engines.
2. Most military aircraft have piston engines.
3. Light aircraft usually have one or two jet engines.
4. Helicopters always have piston engines.
5. Piston engines produce power by burning petrol and air.
6. Jet engines have turbines and cylinders.

Language Point

> Some helicopters have jet engines.
> Some have piston engines.
> Most light aircraft have one or two engines.

Practice 2

Look at this example:
Most military aircraft have ... *four* ... engines.

Now complete these sentences in the same way.

1 Most light aircraft have .... engines.
2 Most aircraft piston engines have four or eight ....
3 Most cars have four ....
4 ... calculators have sixteen or twenty ....
5 .... cars .... headlights.
6 .... windscreen wipers.
7 .... wings.
8 ....

Practice 3

Look at this example:
Some helicopters have piston engines. *Some have jet engines.*

Now form sentences in the same way.

1 Some commercial aircraft have three engines. ....
2 Some light aircraft have one engine. ....
3 Some helicopters are commercial. ....
4 Some engines burn petrol. ....
5 Some lorries have six wheels. ....
6 ....

Study Section 1.2

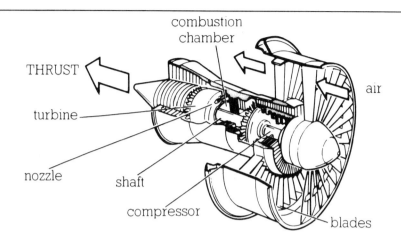

The construction of a jet engine is quite simple. The turbine and compressor blades are connected to a shaft. The shaft rotates at high speed (up to 20,000 rpm). The blades of the compressor compress air and pass it at high speed into the combustion chamber. A nozzle sprays fuel into the combustion chamber. The fuel and air mixture burns rapidly and expands. The burning mixture passes over the turbine blades. The turbine rotates and turns the compressor. The burning gases leave the engine at very high speed. This produces thrust.

Temperatures inside the turbine and combustion chamber are high – up to 2,000 °C in some engines.

Practice 4

These sentences contain mistakes. Find the mistakes.

1. Turbines have blades and compressors have shafts.
2. Jet engines rotate at speeds up to 2,000 °C.
3. The compressor burns the air and fuel.
4. The fuel sprays the nozzle into the combustion chamber.
5. The compressor turns the turbine.
6. The turbine is at the front of the engine.

Language Point

1 m	one metre
10 m	ten metres
1 °C	one degree centigrade
10 °C	ten degrees centigrade

Practice 5

Look at this table:

mm	millimetre(s)
cm	centimetre(s)
m	metre(s)
km	kilometre(s)
°	degree(s)
rpm	revolution(s) per minute

Now read these out:

1. 1 cm
2. 12 mm
3. 1 km
4. 50 m
5. 32 °C
6. 100 °F
7. 1 °F
8. 5,000 rpm
9. 75 kph
10. 1,000 kph

Is 0 °C the same as 0 °F?
Is 150 mm longer than 10 cm?

Practice 6

Look at this diagram of a piston engine:

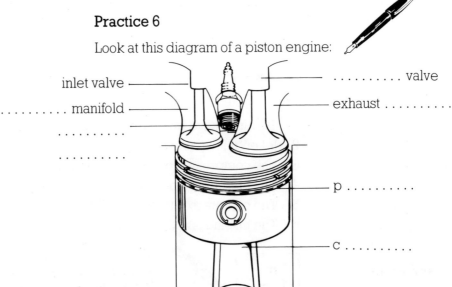

inlet valve
.... valve
.... manifold
exhaust ....
....
....
p ....
c ....

1. Label the diagram.
2. Look at the diagram of a jet engine on page 3 and discuss the differences between piston engines and jet engines.

Practice 7

Look at this example:

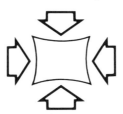

..........*compress*..........

1 Now describe these in the same way:

a)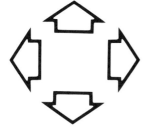

b)

c)

....

d)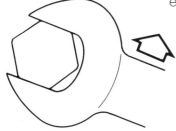

e)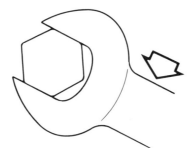

f)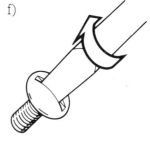

....

2 What are the tools in pictures c, d, e and f above?
3 What are they used for?

Practice 8

Find out the differences between the rotational speeds of:

1 jet engines
2 piston engines used in cars
3 piston engines used in aircraft
4 piston engines used in ships

Study Section 1.3

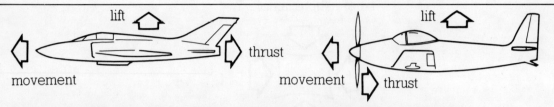

Jet engines produce thrust. Piston engines in aircraft are connected to a propeller. Propellers also produce thrust.

Thrust produces movement. The wings of an aircraft produce lift. The control surfaces of an aircraft produce lateral, longitudinal and rotational movement.

The control surfaces of an aircraft

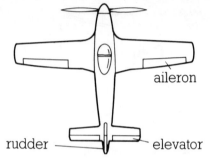

The elevators produce longitudinal movement (upwards/downwards):

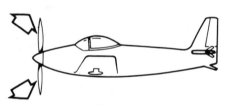

The rudder produces lateral movement (to the left/right):

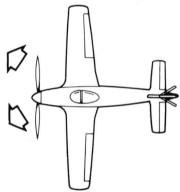

The ailerons produce rotational movement (clockwise/anti-clockwise):

The control surfaces are connected to controls inside the aircraft. The pilot moves the controls and moves the aircraft. In most light aircraft the connection between the controls and the control surfaces is mechanical. In most military and commerical aircraft it is hydraulic.

Practice 9

Label this diagram:

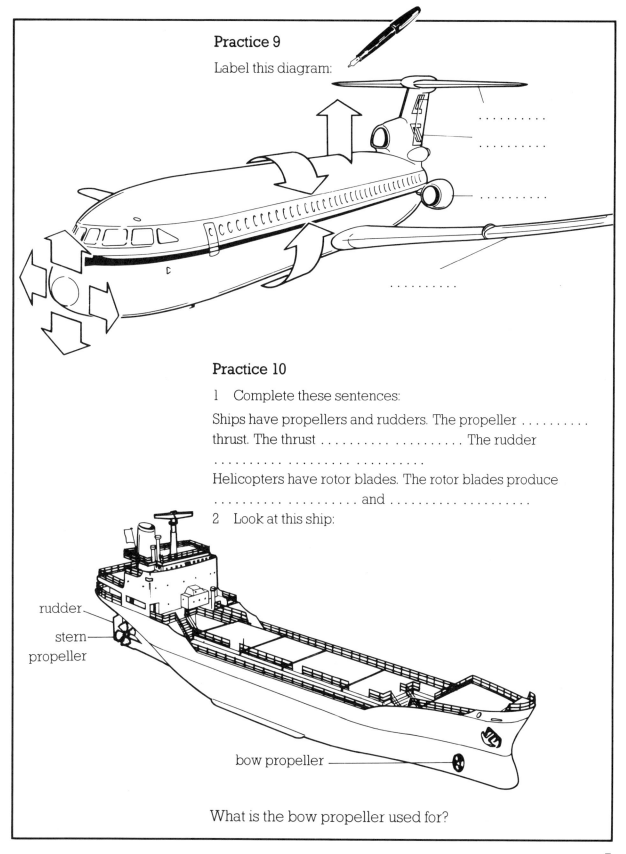

Practice 10

1 Complete these sentences:

Ships have propellers and rudders. The propeller .... thrust. The thrust .... The rudder ....
Helicopters have rotor blades. The rotor blades produce .... and ....

2 Look at this ship:

What is the bow propeller used for?

Practice 11

1. Describe these three different connections using the words below:

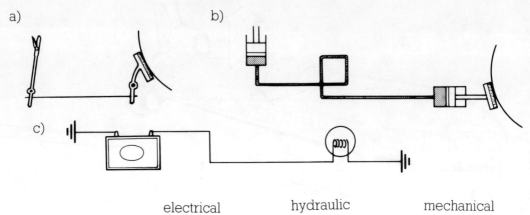

electrical   hydraulic   mechanical

2. Some parts of a car are mechanical. Some parts are electrical. Some are hydraulic. Which parts are mechanical? Which are electrical? Which are hydraulic?
3. Look at this diagram of a car:

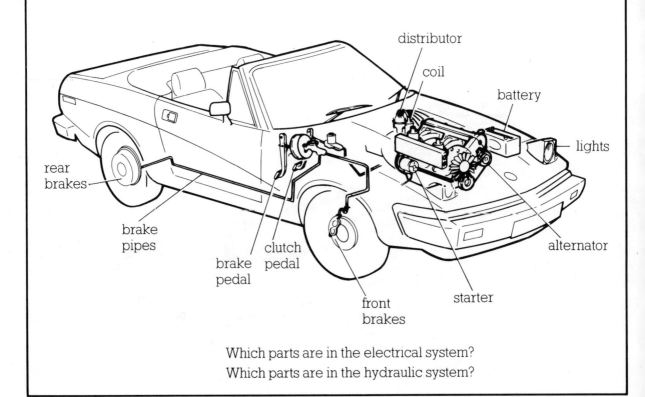

Which parts are in the electrical system?
Which parts are in the hydraulic system?

Practice 12

Look at these diagrams and decide which way the vehicles will move. (Upwards/downwards, to the left/right, clockwise/anti-clockwise.)

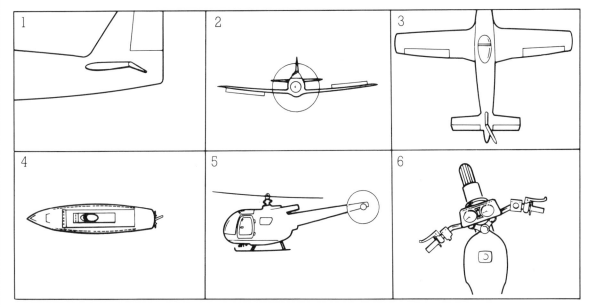

Practice 13

Look at this example:
An aircraft has a ... *pilot* ...

Find words to put in these sentences:

1 A car has a ....
2 A motorcycle has a ....
3 A ship has a ....
4 A helicopter has a ....
5 A bicycle has a ....
6 A lorry has a ....

Word List

an aircraft
an aileron
a blade
a combustion chamber
a compressor
a construction
a control surface
a cylinder
an elevator
a helicopter
a jet engine
a nozzle
a piston engine
a propellor
a rotor blade
a rudder
a shaft
a turbine

construction
fuel
kerosene
lift
thrust

commercial
electrical
hydraulic
lateral
longitudinal
mechanical
military
rotational

rapidly

up to ...

burn
compress
expand
produce
rotate
spray

some helicopters have ...
some have ...

Unit 2
Levers and Machines

Study Section 2.1

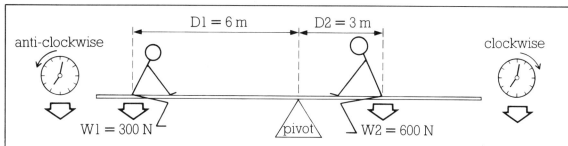

This see-saw is balanced. It does not move. There is a boy on the left of the pivot and a man on the right.

The boy exerts a force of 300 N (his weight) and is six metres from the pivot. The man exerts a force of 600 N and is three metres from the pivot. There is no movement because the man and the boy exert the same turning force. This turning force is usually called a moment.

The turning force around a pivot is equal to the force times its distance from the pivot:

moment of force = force × distance from pivot
        = weight × distance from pivot of boy
        = W1 × D1
        = 300 N × 6 m
        = 1800 Nm (Newton metres)

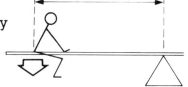

moment of force = W2 × D2
        = 600 N × 3 m
        = 1800 Nm

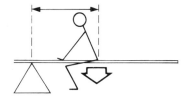

The boy exerts a turning force in an anti-clockwise direction. This moment is equal to the force multiplied by the distance and is equal to 1800 Nm. The man exerts the same turning force in a clockwise direction. Why is there no movement? Because the moments are equal.

What happens if the man moves away from the pivot?

If the man moves away from the pivot D2 increases and the clockwise moment also increases. The see-saw rotates in a clockwise direction.

What happens if the boy moves towards the pivot?

Practice 1

Are these statements true or false?
Correct the false ones.

1. The see-saw is balanced because $D1 \times W1 = D2 \times W2$.
2. The boy exerts a force of six metres.
3. The moment of a force is the force multiplied by the distance from the pivot.
4. Five times six is eleven.
5. If the man moves away from the pivot the see-saw rotates in a clockwise direction.
6. If the boy moves away from the pivot the see-saw rotates in a clockwise direction.
7. If the boy moves away from the pivot, D1 increases.
8. The moment of a force is measured in Newtons.

Practice 2

Look at this simple balance:

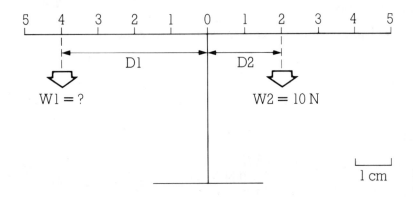

Make statements about the balance.

For example:
If W1 = 5 N, the balance .... *does not move* ....

Use

moves	in	a clockwise an anti-clockwise	direction
does not move			

1 If W1 = 4, ...
2 If W1 = 10, ...
3 If W1 = 5, ...
4 If W1 = 5 and D1 = ... clockwise ....
5 If W1 = 5 and D1 increases, ...
6 If W1 = 5, D1 = 4 and W2 = 1 and D2 = 20, ...
7 ....
8 ....

Language Point

> If W1 *increases*, the see-saw *rotates*.

Practice 3

Look at this language point:

Language Point

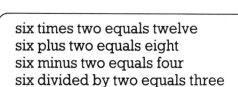

six times two equals twelve
six plus two equals eight
six minus two equals four
six divided by two equals three

Now read out these calculations.

1 $D1 \times D2 = 20 \text{ m}^2$
2 $T \times 5 = 22\,°C$
3 $9 - 3.1 = ?$
4 $W \times D = 16 \text{ Nm}$
5 $2.5 + 2.6 = ?$
6 $56 \div 7 = ?$
7 $? \times 6 = 18$
8 $27\,°C + 38\,°C = ?$
9 $7.1 \times 2.3 = ?$
10 $3.3 \div 1.1 = ?$
11 $7 + 26 = ?$
12 $\frac{1}{2} + \frac{1}{4} = ?$
13 $W2 \times D2 = 33.02 \text{ Nm}$
14 $81 \div 3 = ?$
15 $57 - ? = 31$

Study Section 2.2

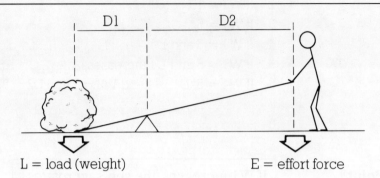

L = load (weight)　　　　　　　E = effort force

This is a lever. The man presses on the lever. He applies a force, an effort force. If D2 × E is greater than L × D1, the lever rotates in a clockwise direction and the rock moves. The effort force is less than the weight of the rock because D2 is greater than D1.

The lever helps the man. It is a simple machine. Levers are used for moving big loads with smaller effort forces. Levers magnify the effort. When a lever mangifies an effort force five times, its *mechanical advantage* is five.

Mechanical advantage = $\dfrac{\text{load}}{\text{effort}}$

$MA = \dfrac{L}{E}$

Practice 4

Are these statements true or false?
Correct the false ones.

1 A lever is a machine.
2 The man applies a load force.
3 The weight of the rock is the same as the load force.
4 A machine uses big efforts for moving small loads.
5 When an effort force of 5 N moves a load of 50 N, the mechanical advantage of the lever is 25.
6 When a lever magnifies an effort force seven times, the mechanical advantage is seven.
7 $MA = \dfrac{E}{L}$
8 A spanner is a machine.

Language Point

E is greater than L.
D1 is less than D2.
MA equals L over E.

Practice 5

Read these out:

1. $L > D$
2. $W_1 < W_2$
3. $V = \dfrac{d}{t}$
4. $p = \dfrac{q}{v}$
5. $D < 5.21\,m$
6. When $D_1 > D_2$, $E = L$.
7. $x = y^2$
8. When $L = 5$ and $E = 4$, $MA > 1$.

Practice 6

Make statements about these levers.

For example:

1
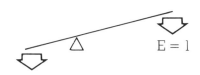

When L equals 5 and E equals 1 the mechanical advantage is 5.

2
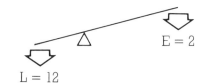

....
....

3

4
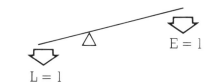

....
....

Study Section 2.3

There are three types of lever.

Type 1

These levers have the pivot between the effort and the load. A small effort moves a large load.

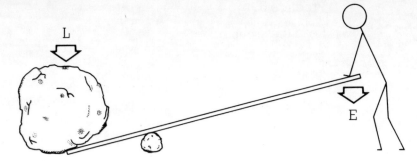

Type 2

These levers have the load between the effort and the pivot. The effort force is magnified.

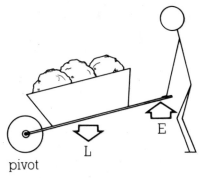

Type 3

These levers have the effort between the load and the pivot. The effort is greater than the load.

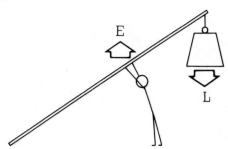

Practice 7

Look at these examples of levers. Label the effort, the load and the pivot. Write a few sentences about each lever.

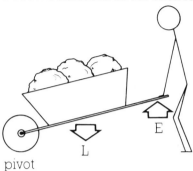

This is an example of the second type of lever. The load is between the effort and the pivot. The effort force is magnified.

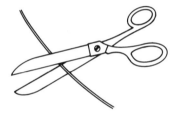

This is an example of the ....
type of lever. The .... is
between the .... and the
.... A small effort moves
....

This is .... the
.... of lever. The
....
.... and the
.... The effort
force ....

....
....
....
....
....

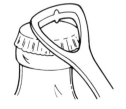

....
....
....
....
....

Practice 8

1 You want to loosen a very tight nut. Which spanner do you use? A or B? Why?

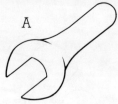
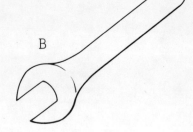

2 What is his mistake?

3 Hold a ruler in your left hand. Place a weight on the ruler near your hand. Move it along the ruler. What happens and why?

Practice 9

What happens if . . . .?
Write correct sentences from the information in columns A and B.

A	B
A fuel and air mixture burns.	contract
Water is heated to 100 °C.	expand
A load is attached to a spring balance.	stretch
We use a 3 A fuse with a cooker.	melt
Burning gases leave a jet engine at speed.	boil
We move the rudder of a boat.	produce thrust
We cool metal.	produce lateral movement

For example:
If a fuel and air mixture burns, it expands.

1 ....
2 ....
3 ....
4 ....
5 ....
6 ....
7 ....

Practice 10

Look at this metre rule. It is balanced on a screw.

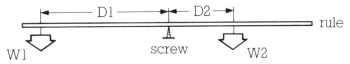

Different weights are used for balancing the ruler.
The ruler **always** balances.
Look at this table. Complete the results.

W1 N	D1 m	Anti-clockwise moment Nm	W2 N	D2 m	Clockwise moment Nm
5	0.5			0.5	
5	0.2		10		
3	0.1		6		
2		0.6	3		

Practice 11

Make statements about the results of Practice 10.

For example:
When W1 = 5 N and D1 = 0.5 m, the anti-clockwise moment is 2.5 Nm.

Practice 12

Think of examples of levers in the classroom and in your home.
Draw diagrams of them and label the diagrams.
Write sentences about them.

Word List

an effort force
a force
a lever
a load
a mechanical advantage
a moment
a Newton metre
a pivot
a see-saw
a turning force

anti-clockwise
balanced
clockwise
equal

divided by
minus
plus
times

apply
equal
exert
magnify

... is greater than ...
... is less than ...
if/when W1 = 5 N the balance does not move

Unit 3
Engineering Materials

Study Section 3.1

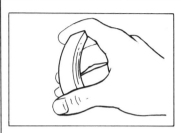
An eraser is made of rubber. Rubber is an engineering material. If you bend it between your fingers it changes shape. When you release it, it regains its original shape. Rubber is very elastic. Elasticity is a property of some engineering materials.

If you hit a piece of glass it breaks. Glass is very brittle. Brittleness is a property of glass.

Nails are made of a tough material. If you hit one with a hammer it doesn't break.

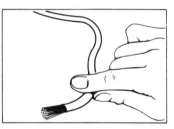
Electrical wires are covered with plastic. Plastic is a bad conductor of electricity. If you touch the plastic you don't receive an electric shock. The wire is made of copper. Copper is a good conductor of electricity. Plastic, however, is an insulator. Copper is also a good conductor of heat.

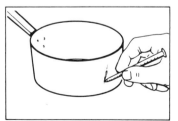
You cannot scratch glass easily. However, if you scratch an aluminium saucepan with a nail it leaves a mark. Glass is a hard material but aluminium is quite soft.

Practice 1

Are these statements true or false?
Correct the false ones.

1. Rubber is a very brittle material.
2. If you strike a brittle material, it doesn't break.
3. Plastic is a good conductor of electricity.
4. When you hit a tough material with a hammer, it breaks easily.
5. A spring is elastic.
6. Glass is soft and brittle.
7. Some saucepans are made of copper because it is a good conductor of electricity.
8. An elastic material changes its shape easily.

Practice 2

Look at this example:
If you hammer a .. *brittle* .. material, it breaks easily.

Now complete these statements:

1. Elasticity is a .... of some engineering ....
2. .... is a property of glass.
3. Copper is a .... of electricity.
4. If you stretch a piece of rubber it ....
5. When you release the rubber it .... its ....
6. If ....
   ...., it doesn't break easily.

Study Section 3.2

Look at this table. It compares the properties of five engineering materials:

HEAVY	TOUGH	HARD	A GOOD CONDUCTOR
copper cast iron glass aluminium rubber	copper aluminium rubber cast iron glass	glass cast iron copper aluminium rubber	copper aluminium cast iron glass rubber
LIGHT	WEAK	SOFT	A POOR CONDUCTOR

Copper is a tough material. It is tougher than rubber.

Aluminium is a light material. It is lighter than glass.

Copper is a good conductor of electricity. It is a better conductor than aluminium.

Cast iron is a heavy material. It is heavier than aluminium.

Rubber is a bad conductor of electricity. It is a worse conductor than cast iron.

Glass is a very brittle material. It is more brittle than rubber.

Practice 3

Make true statements about the table.

Look at these examples:
1. Is cast iron lighter than glass?
   No it isn't. It's heavier.
2. Aluminium is a better conductor of electricity than cast iron, but a worse conductor than copper.
3. Copper is a heavy material. Rubber, however, is light.

Language Point

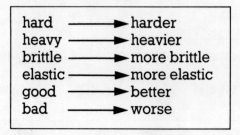

Practice 4

Compare the properties of these engineering materials. Write sentences from the table.

materials	properties
copper	tough
cast iron	hard
silver	soft
rubber	light
aluminium	brittle
glass	a good conductor of { electricity, heat }

Look at the example:
Cast iron is heavier than rubber. It is also a better conductor of electricity.

Practice 5

Compare the dimensions of these different objects.

For example:

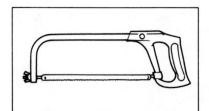 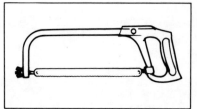

The first hacksaw has a narrower blade than the second one.

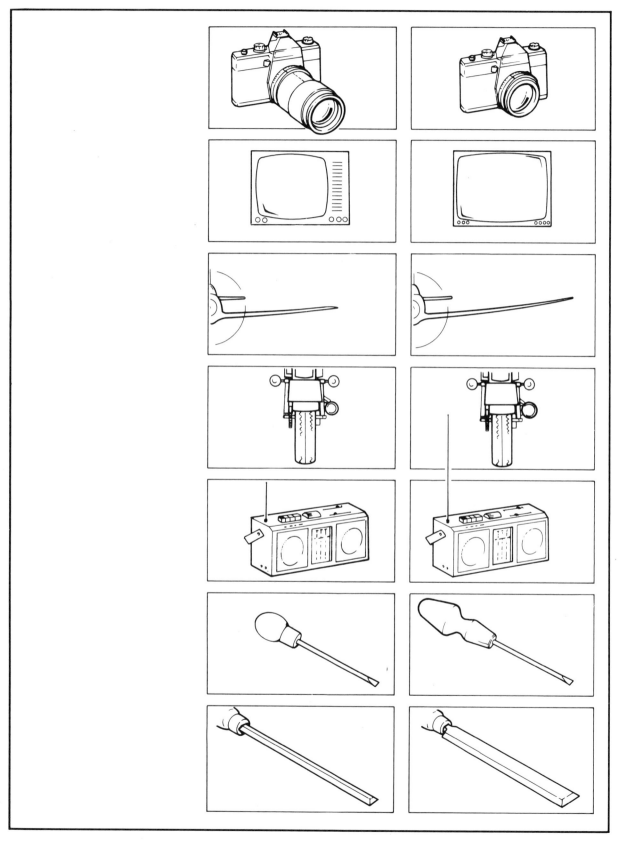

25

Practice 6

Examine some of the objects in your classroom and your home. Make notes about their properties and dimensions.

object	material	dimensions	properties
saucepan	aluminium	20 cm wide 17 cm high	light a good conductor of heat

Now make statements about the table.
For example:
The saucepan is made of aluminium. It is approximately 20 cm wide and 17 cm high. It is light and a good conductor of heat.
Now compare the dimensions and properties of the different objects.

Study Section 3.3

1. Bolts are made of mild steel, which is a type of ferrous metal. All ferrous metals contain iron. All steels contain iron and a small quantity of carbon. If the quantity of carbon changes the properties of the steel change.

Look at this table. It shows the quantities of carbon in different types of steel.

	approximate carbon content
wrought iron	0.05%
mild steel	0.08% – 0.25%
medium carbon steel	0.25% – 0.65%
high carbon steel	0.65% – 1.5%
tool steel	1.05%

2. Wrought iron is weaker than mild steel and it is more expensive to produce. However, it is very tough and is therefore used for making crane hooks. Mild steel is less expensive than wrought iron because it is easier to produce. It is used in the
5 manufacture of girders and nuts and bolts.
Medium carbon steel contains more carbon than mild steel and is tougher and harder. It is also more expensive to produce. It is used for making screwdrivers, spanners, hammer heads etc.
High carbon steel contains more carbon than medium carbon
10 steel and is therefore harder and tougher. It is used in the manufacture of knives, axe-heads and other cutting tools.

Some high-speed drill bits, which are very hard, are made of tool steel which contains tungsten. Tool steel is more expensive than the other steels.

Practice 7

Look at this sentence from Study Section 3.3:
Wrought iron is weaker than mild steel and it is more expensive to produce.

In this sentence *it* means wrought iron. What does *it* refer to in:

1 line 2?
2 line 4?
3 line 4?
4 line 7?
5 line 7?
6 line 10?

Practice 8

Are these statements true or false?
Correct the false ones.

1 Mild steel is stronger than wrought iron.
2 Medium carbon steel contains more carbon than high carbon steel.
3 High carbon steel is harder and tougher than medium carbon steel because it contains more carbon.
4 Girders are made of mild steel.
5 Cutting tools are made of steels which are very hard.
6 Mild steel is less expensive than wrought iron because it contains more carbon.
7 Knives are made of tool steel.
8 Mild steel is cheaper than tool steel.

Language Point

6,801
0.6801

six thousand eight hundred and one

zero point six eight zero one

Practice 9

Read these out:

1. 68.068%
2. 1.025 cm^2
3. 47.037 Nm
4. 59.591 m
5. 12.125 mm
6. 38.56 °C

Practice 10

Name the objects below.
Look at this example:
Bolts are made of mild steel.
Mild steel contains 0.08 – 0.25% carbon.

Now describe these in the same way.

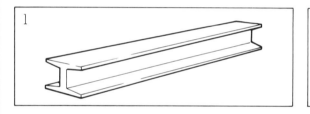
....

....

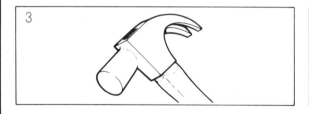
....

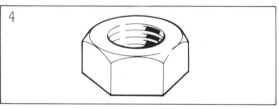
....

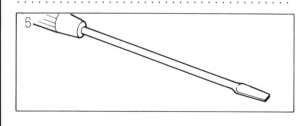
....

....

Practice 11

Look at this example:
Mild steel is a type of ferrous metal.
Mild steel contains 0.08 – 0.25% carbon.
Mild steel is a type of ferrous metal which contains 0.08 – 0.25% carbon.

Now make similar sentences with your answers from Practice 10.

Practice 12

Write sentences using *therefore*.

Look at the example:

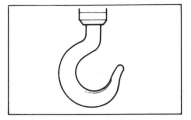

Wrought iron is very tough and is therefore used for making crane hooks.

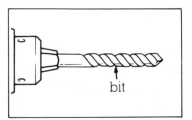

Tool steel ....
....
....

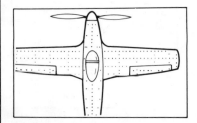

Aluminium ....
....
....

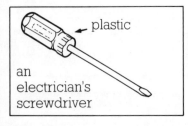

Plastic is a bad ....
....
....

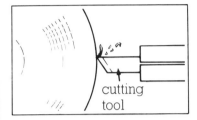

Plastic is a good ....
....
....

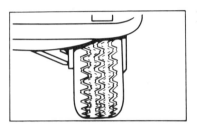

Tool steel ....
....
....

Rubber ....
....
....

Plastic ....
....
....

Now rewrite the statements using *because*.

For example:
Wrought iron is very tough and is therefore used for making crane hooks.
Crane hooks are made of wrought iron because it is very tough.

Practice 13

1. Find out the melting points of copper, zinc, aluminium, cast iron and any other materials. Put the results in a table. Compare the different melting points.
2. Compare any two motor cars. Find out about the dimensions, engine size, price, etc. Put the results in a table. Compare the results.

Word List

	a bolt	brittle
	a conductor	elastic
	a crane hook	electrical
	a cutting tool	ferrous
	an engineering material	hard
	an eraser	heavy
	a girder	light
	an insulator	mild
	a nut	soft
	a property	strong
	a rubber	tough
	a tungsten	weak

brittleness
carbon
cast iron
copper
elasticity
ferrous metal
high carbon steel
iron
medium carbon steel
metal
mild steel
quantity
steel
tool steel
tungsten
wrought iron

made of

because
however
therefore
which

change shape
contain
regain
scratch

bad-worse
good-better
light-lighter than
. . . a metal which contains . . .

Unit 4
Electricity

Study Section 4.1

Copper is a good conductor of electricity. If a copper wire is connected to a battery, an electric current flows in the wire. However, a current does not flow if you connect a piece of plastic to the battery. Because plastic is an insulating material, it resists the flow of electricity. Bad conductors of electricity have a high resistance.

Look at this diagram. It shows an electric circuit.

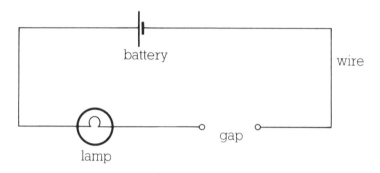

Place a piece of copper wire across the gap. The lamp lights. Current flows around the circuit. After removing the copper, place a piece of nichrome across the gap. The lamp is dimmer. Nichrome has a high resistance and the current does not flow easily.

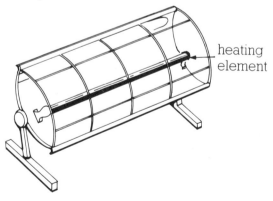

This fire uses electricity for heating. The heating element is made of nichrome which is a bad conductor of electricity. It has a high resistance and current cannot flow easily. Therefore, the current passing through the element generates heat.

Practice 1

Are these statements true or false?
Correct the false ones.

1. An insulator is made of a material which resists the flow of electricity.
2. An electric current flows easily through a good conductor.
3. If a copper wire is connected to a battery an electric current flows.
4. Nichrome is a good conductor of electricity.
5. If you pass an electric current through nichrome, heat is generated.
6. Rubber has a low resistance to electric current.
7. An electric current flows easily through a material which has a high resistance.
8. The lamp is dimmer when a smaller current flows.

Practice 2

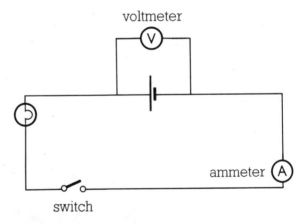

Make statements about the symbols.
For example:

This is the symbol used for a lamp in an electric circuit diagram.

1 2 3 4 5

(An ammeter is used for measuring electric current.
A voltmeter is used for measuring electrical energy.)
Now look at the circuit diagram on page 34 and read this:
The battery is in the middle of the diagram at the top and a voltmeter is connected across it. The lamp is on the left, towards the top, and the ammeter is on the right, towards the bottom. The switch is positioned at the bottom and on the left.

Now read this and draw the circuit diagram which it describes. The battery is in the middle of the diagram at the bottom and a voltmeter is connected across it. On the right of the diagram is a lamp which is positioned towards the top. The ammeter is on the left and in the middle and the switch is at the top on the right.

Practice 3

Look at this circuit diagram and complete the description.

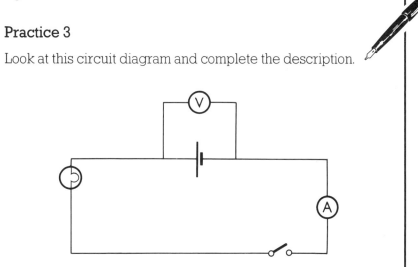

The battery .... in the .... of the .... at the .... and a .... is .... it.
The .... is in .... on .... right. The lamp is on .... towards .... The switch ....

Language Point

> Resistance (R) is measured in ohms (Ω)
> Current   (I) is measured in amps (A)
> Voltage   (V) is measured in volts (V)

Practice 4

Make statements like the example:

An ammeter is used for measuring current.
Current is measured in amps.

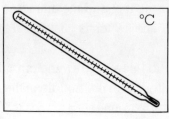

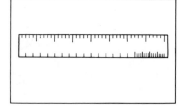

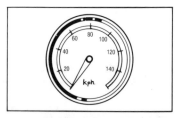

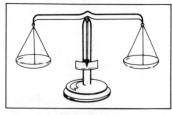

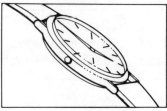

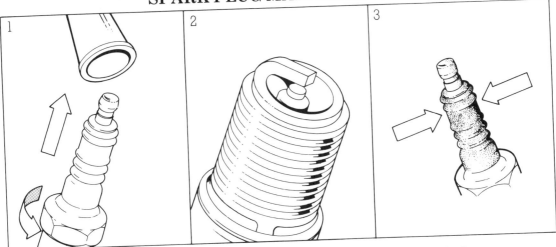

SPARK PLUG MAINTENANCE

First remove the plug lead. Then unscrew the spark plug. (1) Check the condition of the electrodes. (2) Clean them with emery paper. Measure and adjust the gap between them. Clean any oil or grease from the outside insulation. (3) (Use petrol or Kerosene.) Replace the spark plug in the engine. (Do not overtighten.) Replace the lead. Repeat this for all plugs. Test the engine.

Look at this example:

Before removing the spark plug, remove the lead.
After removing the lead, remove the spark plug.

Now complete these in the same way:

1 After checking the condition of the electrodes, ...
2 Before measuring and adjusting the gap between the electrodes, ...
3 After .... the electrodes with emery paper, ...
4 After .... the outside insulation, ...
5 Before .... the lead, ...
6 ...., test the engine.

We write:
Remove the plug lead.

We usually say:

> You should remove the plug lead.

Complete these:

1. Remove the spark plug.

   > ....

2. ....
   ....

   > You should check the condition of the electrodes.

3. After adjusting the gap between the electrodes, clean them with emery paper.

   > After adjusting the gap between the electrodes, ...

4. Before replacing the spark plug in the engine, ...
   ....

   > ....
   > ....
   > ....

5. After replacing the spark plug in the engine, ...
   ....

   > ....
   > ....
   > ....

6. Before testing the engine, ...
   ....

   > ....
   > ....
   > ....

Language Point

loosen	loosening
tighten	tightening
remove	removing
secure	securing
put	putting
fit	fitting

Practice 6

This formula is used to calculate the resistance in a circuit.

$$V = I \times R \text{ (V equals I times R)}$$

Fill in the gaps:
V = voltage measured in volts, V.
I = ...
R = ...

Look at this circuit diagram:
─▭─ is the symbol for a resistor. A resistor is made of a material which resists the flow of electricity. Some resistors are made of nichrome.

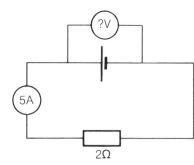

To calculate the voltage, V:
$V = I \times R$
$= 5 \times 2$
$= 10$
$V = 10$ volts

Read out the calculations in full, for example, *voltage equals current times resistance, which is five times two . . .*

Now calculate the current, voltage and resistance in the following circuits:

Calculate I   Calculate V   Calculate R

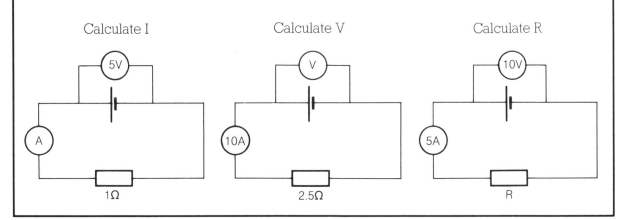

Practice 7

Put in the right words. The first one has been done for you.

1. If you ..*connect*.. a .... wire to a battery, a .... flows.
2. After .... the copper, place a piece of nichrome across the ....
3. An insulator .... the flow of electricity.
4. An .... is used for .... current.
5. .... is measured in ohms.
6. A voltmeter is connected .... a battery.
7. The switch is .... the top .... the right.
8. Draw a .... diagram which has a battery on the right.
9. Heat is .... when you pass an electric current .... nichrome.
10. To calculate the voltage use this ....

Practice 8

1. Look at the batteries in a calculator, a torch and a cassette player. Write down the voltages.
2. If the torch uses 0.5 A, the calculator 0.05 A and the cassette player 2 A, calculate the resistances.
3. On the back of most electrical devices there is the power rating:

```
SONIO JAPAN
MODEL 36K
SERIAL No 0169423/005
6W 6V
```

Use this formula for calculating the power of an electrical device:

$$P = I \times V$$

Power, P, is measured in Watts, W. The power rating of the device is 6W. Calculate the current which is used:

$$P = I \times V$$
$$6 = I \times V$$
$$\frac{6}{6} = I$$
$$I = 1 A$$

4   Look at this table. Fill in the gaps.

electrical device	power (W)	voltage (V)	current (A)
electric fire	2,400 W	240 V	
light bulb	55 W	110 V	
electric drill	990 W	110 V	
electric kettle		240 V	10 A
car headlight	48 W	12 V	
television	220 W	110 V	
cassette player		6 V	1 A

5   Make true statements.
Here are some examples:
The electric kettle has a power rating of 2,400 A.
The electric fire is connected to a 240-volt mains supply.
The cassette player draws a current of 1 A.

Now make statements like this example:
The electric drill has a power rating of 990 W. When it is connected to a mains supply of 110 volts it uses a current of 9 A.

Practice 9

Fuses usually have four different ratings:
3 A, 5 A, 10 A, 13 A
Make statements about the electrical devices in Practice 8 like this example:
Use a 13 A fuse with an electrical kettle because it draws a current of 10 A.

Practice 10

1   Calculate the resistance of the electrical devices in Practice 8.
2   Find out the power rating of other electrical devices, for example, an electric cooker, washing machine, radio etc.

Language Point

240 V – two hundred and forty volts
50 A – fifty amps

Study Section 4.2

At some power stations, coal is used for generating electricity. Coal is placed on a conveyor belt which tips it into a bunker. The coal leaves the bunker and is ground into a powder. The coal is then blown into a furnace and it burns.

The furnace heats water in the boiler tubes and steam is produced at a very high temperature. The steam is then heated and passes to the turbine. The turbine rotates and turns the generator. A high voltage is generated. The steam is cooled and passes back to the boiler.

Practice 11

Label this diagram:

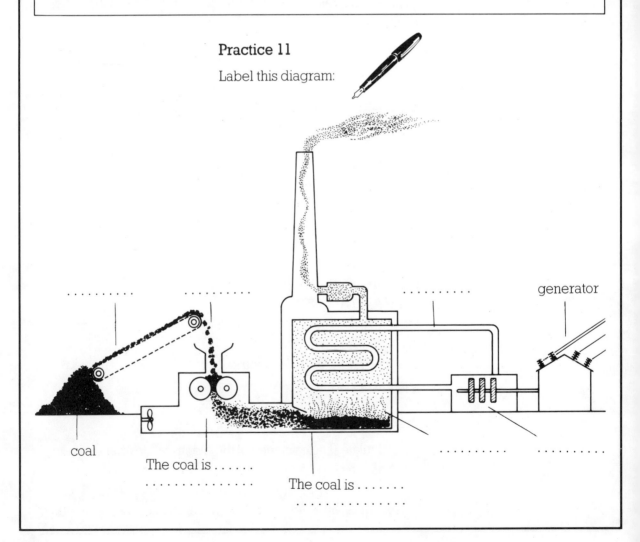

generator

coal

The coal is ......
....

The coal is ....
....

Practice 12

Are these statements true or false?
Correct the false ones.

1. The conveyor belt tips the coal into the bunker.
2. The coal burns in the furnace.
3. Most power stations use coal.
4. The coal is blown into the furnace before leaving the bunker.
5. The steam rotates the turbine.
6. Power stations generate high voltage electricity.
7. The steam is heated and then led to the generator.
8. Electricity is generated because the turbine rotates.

Practice 13

Choose the right form of the verb in brackets. Then put the sentences in the right order.

For example:
Most power stations (use) coal.
Most power stations use coal.
Coal (use) for generating electricity.
Coal is used for generating electricity.

1. The turbine (rotate) and (turn) the generator.
2. The coal (leave) the bunker and (grind) into a powder.
3. The furnace (heat) the water and steam (produce) at a high temperature.
4. Coal (place) on a conveyor belt which (tip) it into a bunker.
5. A high voltage (generate).
6. The steam (heat) and (pass) to the turbine.
7. The coal (blow) into a furnace and it (burn).
8. The steam then (cool) and (pass) back to the boiler.

Language Point

connect-id
heat-id
generate-id

connected
heated
generated

cooled
removed
loosened

placed
produced
compressed

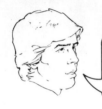

place-t
produce-t
compress-t

cool-d
remove-d
loosen-d

Practice 14

Make true sentences from the table.

Coal Water Steam Electricity	is	heated tipped placed ground blown generated cooled	on into in by	a conveyor belt. the bunker. a furnace. a powder. the boiler. the generator.
		leaves burns		

Now look at this:
After leaving the bunker, the coal is ground into a powder.
or
Before being ground into a powder, the coal leaves the bunker.

Language Point

leave~~e~~   leaving
remove~~e~~   removing

is ground   being ground
is heated   being heated

Word List

an ammeter	electric current	
an amp	emery paper	
a battery	electrical energy	
a boiler tube	flow	
a bunker	insulation	
a circuit	nichrome	
a conveyor belt	resistance	
a device	voltage	
an electrode		
a formula	dim	
a furnace	adjust	
a gap	calculate	
a generator	cool	
a heating element	generate	
an insulator	grind	
a main supply	heat	
an ohm	measure	
a power rating	overtighten	
a resistor	pass through	
a switch	resist	
a volt		
a voltmeter	after/before removing . . .	
a watt	. . . you should remove . . .	

Unit 5
Radio and Television

Study Section 5.1

A vibrating guitar string produces a sound.

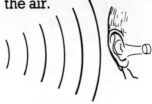

When the string stops vibrating the sound also stops. Sounds are the result of vibrations travelling through the air.

Sounds move through the air in waves.

When a stone is dropped into a bucket of water, ripples radiate from the centre to the side.

Sound waves travel through the air in the same way.

Waves are measured in two ways. Firstly, the wavelength, which is the distance between the tops of two waves:

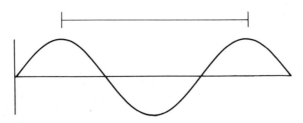

l = wavelength

Secondly, the frequency of the wave, which is the number of waves to reach the side of the bucket in one second:

Wavelength is measured in metres (m). Frequency is measured in cycles per second or Hertz (Hz).

Radio waves are like sound waves but have a higher frequency.

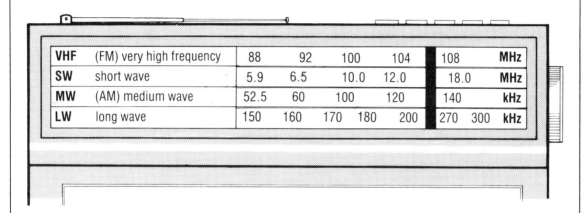

Short radio waves have higher frequencies than long waves. Waves which have a frequency of 200 kHz are approximately 1,500 m long.

Language Point

> MHz = megahertz
> kHz = kilohertz
> M = mega = 1,000,000
> k = kilo = 1,000
> 1Hz = 1 cycle per second

Practice 1

Are these statements true or false?
Correct the false ones.

1. A vibrating guitar string produces a sound wave.
2. Guitar strings are the result of vibrations.
3. Sound moves through the air like water ripples.
4. The length of a wave is the distance between the top and the bottom.
5. If two waves reach the side of the bucket in one second, then the frequency of the waves is 2 Hz.
6. Radio waves with a low frequency are shorter than high frequency waves.
7. Wavelength is measured in cycles per second.
8. All long waves have a frequency of 200 kHz.

Practice 2

Look at these radio stations and their frequencies:

- RADIO 9 — 18MHz
- LONDON RADIO — 180kHz
- RADIO CAPTAIN — 52.5kHz
- RADIO SOUTH — 5.9MHz
- RADIO 5 — 300kHz
- RADIO 10 — 170kHz
- RADIO LOVE — 82.3kHz
- RADIO O.K. — 160kHz
- RADIO 295 — 170.2kHz

Look at the radio in Study Section 5.1 and ask and answer questions like this:

Which frequency does Radio 9 use?

Radio 9 uses 18 Megahertz on short wave.

Practice 3

Put in the correct words.
Choose from these:

therefore, but, the result of, and, however, which, because, when

1. A fuse melting can be .... an increase in electric current.
2. There are commercial helicopters .... there are military helicopters.
3. Aluminium is a tough, light metal and is .... used for making aircraft.
4. The turbine rotates .... turns the compressor.
5. Most piston engines in aircraft have four or eight cylinders .... some have six or twelve.
6. .... the turning forces are equal, the see-saw doesn't move.
7. The heavy load is moved .... the effort is applied at a greater distance from the pivot.

8   Rubber is very elastic. Glass, ..., is very brittle.
9   Most cars are made of steel, ... is a kind of ferrous metal.
10 .... plastic is an insulating material it resists the flow of electricity.
11 The heating element is made of nichrome, ... is a bad conductor of electricity.
12 Sound waves are .... vibrations .... travel through the air.

Practice 4

Complete this table:

measurement of	symbol	unit	abbreviation
frequency	f	cycles per second	Hz
		seconds	s
length	L		
	I		
		volts	
			Ω
		watts	
			°C

Now ask and answer questions and make the answers from the table:

For example:
What are the units of frequency called?
Cycles per second.
or
Frequency is measured in cycles per second.

Practice 5

What are the two formulas for calculating the resistance and power rating of an electrical device?

V =

P =

Now solve these problems:

1. I = 5 A, V = 20 V   calculate P
2. R = 20 Ω, I = 0.5 A   calculate V
3. I = 10 A, V = 10 V   calculate R
4. P = 100 W, I = 10 A   calculate V
5. R = 5 Ω, V = 10 V   calculate I

Read out your answers.

For example: I = 1 A, V = 2 V   calculate R

$$V = I \times R$$
$$R = \frac{V}{I}$$
$$R = \frac{2}{1}$$
$$R = 2\,\Omega$$

You read:

The formula for calculating resistance is V equals I times R. Therefore R equals V over I. If V equals two and I equals one, R equals two over one. Therefore R equals two ohms.

Study Section 5.2

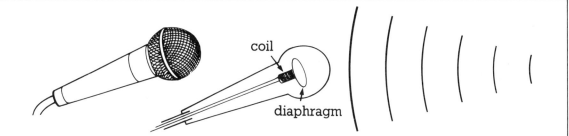

A microphone is like an electric ear. When a sound reaches a microphone it causes a thin piece of metal or plastic to vibrate. This is called the diaphragm. The diaphragm is attached to the coil.

A small electric current is generated by these vibrations. This current flows forwards and backwards very quickly in an alternating cycle.

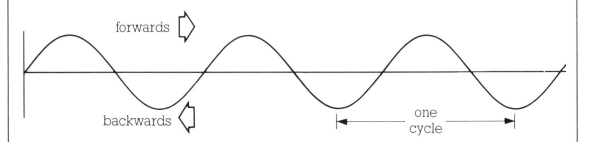

If it does this fifty times in one second, it alternates with a frequency of fifty cycles per second or 50 Hz.

The alternating current in your home usually has a frequency of 50 Hz. A battery produces a current which flows in one direction only. This is called direct current.

The frequency of the alternating current which is generated in the microphone is similar to the frequency of the sound waves. If the alternating current is amplified and connected to loudspeakers, the vibrations which are produced travel through the air as sound waves.

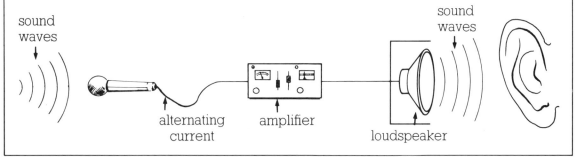

Practice 6

Are these statements true or false?
Correct the false ones.

1. A microphone is an electric ear.
2. The diaphragm is a thick piece of metal or plastic.
3. The alternating current is the result of vibrations of the diaphragm.
4. An alternating current moves forwards in one cycle and backwards in the next.
5. An alternating current which changes direction twenty-five times a second has a frequency of 25 Hz.
6. A torch battery produces an alternating current.
7. The alternating current in a microphone is similar to the sound waves.
8. An amplifier increases the power of the alternating current.

Language Point

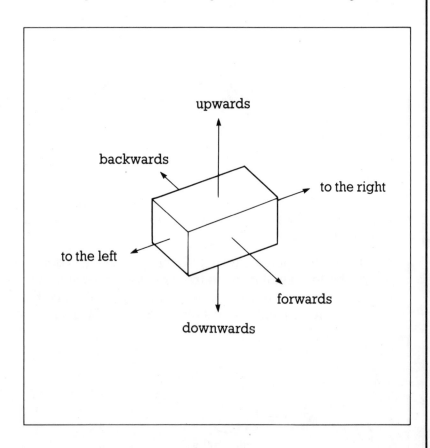

Practice 7

Look at the example and then make instructions.
Move the screw six centimetres to the right.

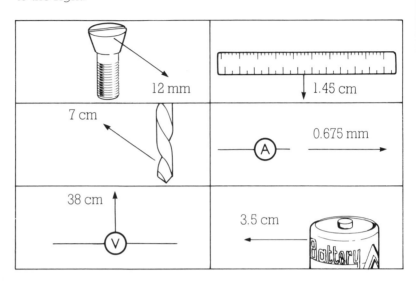

Practice 8

Complete this table with words from this book:

to vibrate	vibration	
	calculation calculator	
	rotation	
to compress		
		moving
to connect		
		elastic
	conductor	
to resist		
	electricity	
		brittle

Now choose the correct word and the correct form:

1. The sounds cause the microphone diaphragm ....
2. Rubber is an .... material.
3. The fuel and air are .... by the piston.
4. The turbine .... and turns the generator.
5. The spark plug leads are .... to the distributor.
6. Aircraft elevators produce longitudinal ....
7. Glass is a very a .... material.
8. An .... current flows through a good conductor very easily.
9. An insulator .... the flow of ....
10. In most light aircraft the .... between the controls and the control surfaces are mechanical.

Language Point

```
apply    – applies
magnify – magnifies
multiply – multiplies
```

Practice 9

Put in the correct form of the verb:

+ TO = to vibrate
+ ING = vibrating
− TO = vibrate
+ S   = vibrates

For example:
When a sound reaches a microphone it causes a thin piece of metal or plastic *to vibrate*. (vibrate)

1. Engines .... power by .... fuel and air. (produce/burn)
2. A nozzle .... fuel into the combustion chamber. (spray)
3. The man .... an effort force against a load force. (apply)
4. An elastic material .... its shape very easily. (change)
5. All ferrous metals .... iron. (contain)

6   Before .... the wheel, ... the wheel nuts. (remove, loosen)
7   Because plastic is an .... material, it .... the flow of electricity. (insulate/resist)
8   After .... the copper wire, ... a piece of nichrome across the gap in the circuit. (remove/place)
9   At some power stations coal is used for .... electricity. (generate)
10 When the clutch .... the engine to the wheels, this causes the car .... (connect/move)
11 Spark plugs .... the fuel in the cylinders by .... a spark. (ignite/make)
12 When the mass equals 1 kg, the balance .... in an anti-clockwise direction. (rotate)

Study Section 5.3

In a radio studio, music and voices travel through the air in sound waves. A microphone generates an alternating current as a result of the vibrations which the sound waves produce. This current is carried into a transmitter and combined with radio waves which are transmitted by a radio transmitting aerial. If one million radio waves leave the aerial in one second they have a frequency of 1 MHz (one megahertz).

When the radio waves reach a receiving aerial on or in a radio they produce a very small alternating current with the same frequency.

Radio waves which have a very high frequency are very short and do not travel a long way. Therefore, most VHF receiving aerials are very big. Medium and long waves are received by an aerial inside the radio.

The aerial receives many radio waves which generate different alternating currents. These currents do not have the same frequency. If you turn the tuning knob you will find the right circuit for the correct wave frequency. The loudspeaker vibrates as a result of these currents and sound waves leave the radio.

Practice 10

Are these statements true or false?
Correct the false ones.

1. Alternating currents which are generated in the microphone combine with radio waves in the transmitter.
2. The transmitting aerial receives radio waves.
3. If 20,000 waves leave the transmitter in one second they have a frequency of 20 MHz.
4. Long waves have a higher frequency than short waves.
5. Some VHF aerials are positioned on the top of houses.
6. Medium waves are received by an aerial inside the radio.
7. The tuning knob is used for finding the circuit for the correct wave frequency.
8. Sound waves are the same as radio waves.

Practice 11

Label the diagram.

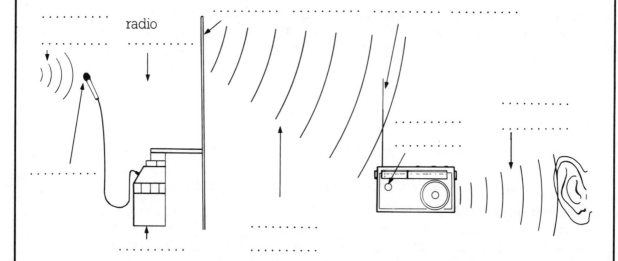

Practice 12

Put the verbs into the correct form.

1. The current (carry) into a transmitter and (combine) with radio waves.
2. Medium and long waves (receive) by an aerial inside the radio.
3. The waves (transmit) by a transmitting aerial.

4 A car battery (produce) a current which (flow) in one direction.
5 When a stone (drop) into a bucket of water ripples (radiate) from the centre to the side.
6 An alternating current (generate) in the microphone.
7 The voltage (increase) from 12 V to 10,000 V by the coil.
8 Aircraft piston engines (connect) to a propeller.
9 The coal (grind) to a powder and (blow) into a furnace.
10 The gearbox of a car (fill) with oil.

Study Section 5.4

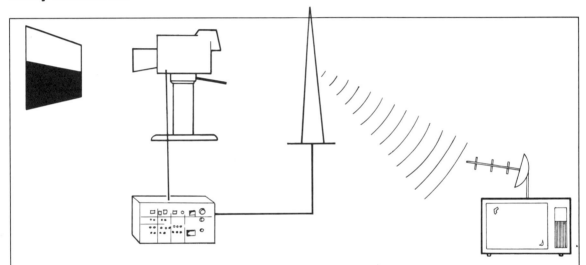

In a radio station, a microphone converts sound waves into small alternating currents. In a television studio the camera converts light into small amounts of electricity called electrons. A bright light generates a large number of electrons, a dim light only a small number.

The electric current is sent to an amplifier and made stronger. Then it is converted into radio waves which radiate from the transmitting aerial.

Because the radio waves travel at the speed of light, the programme is seen and heard at the same time as it is transmitted.

The alternating currents which are produced in the receiving television aerial are carried to the television set. They are converted into direct currents and electrons are fired at the television screen. Lighter parts of the screen are the result of a lot of electrons hitting it, darker parts the result of a few.

The picture which we see is made of hundreds of lines of electrons which hit the screen at very high speed.

Practice 13

Are these statements true or false?
Correct the false ones.

1. Sound waves are converted into small electric currents in a microphone.
2. A television camera converts light into electricity.
3. Small quantities of electricity are called ripples.
4. A dim light generates a large number of electrons.
5. Radio waves are radiated from the transmitting aerial.
6. Radio waves travel at the speed of light.
7. Pictures on a television screen are the result of electrons hitting it.
8. The electrons strike the screen in a number of lines.

Language Point

			countable
a small / a large	number of	electrons units	▨▨▨▨ ▨▨ 1, 2, 3, etc.
	but		
a small / a large	amount of	electricity coal	uncountable 1, 2, 3

Practice 14

Put in the correct words, using the expressions from the Language Point.

1. A radio uses .... electricity.
2. The television screen has light parts as a result of .... electrons hitting it.
3. Large cars usually use .... fuel.
4. A hacksaw blade has .... teeth.
5. A power station generates .... electricity.
6. Mild steel contains .... carbon.
7. A television picture is made of .... lines.
8. A jet engine produces .... thrust.
9. .... hard steels contain tungsten.
10. An electron is .... electricity.

Practice 15

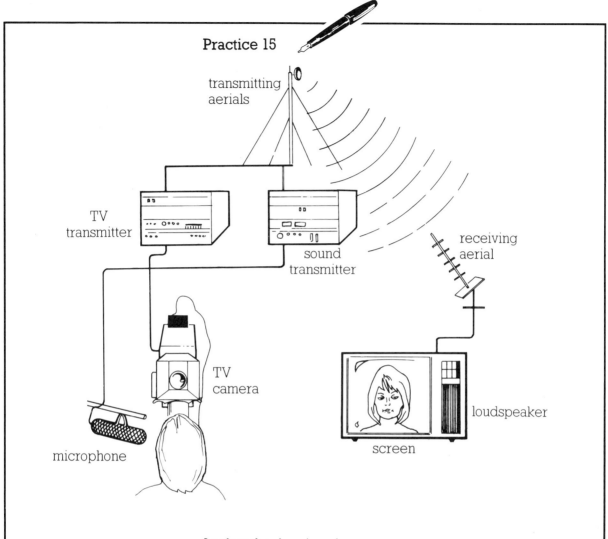

Look at the drawing above.
Write a short description of how a television station and a television receiver operate – transmitting and receiving both sound and pictures.

Word List

an aerial
an alternating cycle
an amplifier
a coil
a diaphragm
an electron
a frequency
a fuse
a guitar
a light wave
a loudspeaker
a microphone
a ripple
a receiving aerial
a result
a screen
a sound wave
a studio
a transmitter
a transmitting aerial
a tuning knob
a vibration
a wave
a wavelength

cycles per second

alternating current
direct current
hertz
kilohertz
megahertz
speed of light

bright

approximately
backwards
downwards
forwards
upwards

to the left/right
the result of

alternate
amplify
combine
convert
melt
radiate
receive
transmit
vibrate

move the screw 6 cm to the right
a small/large number/amount of . . .

Unit 6
The History of Engines

Study Section 6.1

In 1870 a German engineer called Nikolaus Otto designed the first internal combustion engine. The first motor car which used Otto's engine was made in 1875 and Daimler and Benz started selling cars with petrol engines in 1885. Engineers in many countries tried to invent other kinds of engine.

Otto's engine produced power by burning fuel and air. A mixture of petrol and air was compressed and then exploded by a spark. This explosion drove a piston in the cylinder.

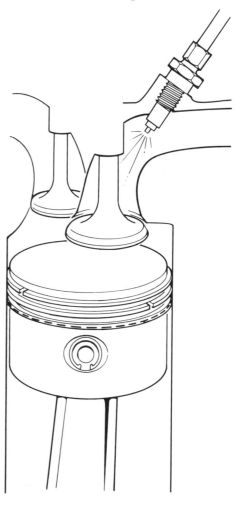

In 1892, however, another German engineer, Rudolph Diesel, created a different type of engine. In the Diesel engine the temperature of the air inside the cylinder was raised to a higher point than in Otto's engine by greater compression. When a fine spray of oil was injected into the cylinder an explosion was caused without a spark.

Diesel's first engine exploded and nearly killed him, but in 1897 he successfully designed and produced his engine.

Diesel's engines were heavier than petrol engines but they had no electrical system or carburettor and they ran on heavier oil.

Nowadays most buses and some taxis have diesel engines. Diesel-electric engines, which are now used on some railway systems, are diesel engines which turn an electric generator. The generator supplies power to an electric motor. Electric motors do not have a gearbox and, combined with a diesel motor, this is very efficient.

Practice 1

Are these statements true or false?
Correct the false ones.
1. Otto designed the first diesel engine.
2. Daimler and Benz designed the first petrol engines.
3. Otto's engine used fuel injection.
4. Diesel raised the temperature of the air inside the cylinder by greater compression.
5. In Diesel's engine internal combustion was caused without ignition.
6. Early Diesel engines were lighter.
7. Most cars use diesel engines.
8. Diesel-electric engines produce power by turning a generator.

Practice 2

Make questions and answers about the history of motor engines.
For example:
1906
What happened in 1906?
In 1906 Charles Rolls and Henry Royce designed the first Rolls Royce motor car.
1 1892   2 1875   3 1870   4 1897   5 1885   6 1906

Practice 3

Internal combustion engines in modern motor cars are like Otto's original engine.

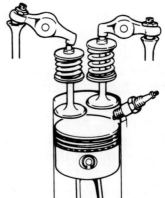

Use these notes to describe the operation of a modern engine, changing the verbs where necessary, like this:
engine – power – burn – fuel – air.
An engine produces power by burning fuel and air.

1. engine – power – burn – fuel – air
2. fuel – air – mix – carburettor
3. inlet valve – open – rocker arm
4. fuel – air – draw into – cylinder – piston
5. then – compress – piston
6. inlet valve – close – spring
7. fuel – air – then ignite – spark plug
8. they burn – expand – quickly – push – piston down
9. exhaust valve – then open – rocker arm
10. burned fuel – air – expel – cylinder – piston

Now describe Otto's 1875 engine.
In 1875 a German engineer called Niklaus Otto designed the first internal combustion engine. It produced power by burning fuel and air. The fuel and air ....
....
....

Now describe Diesel's first engine in the same way. It was like Otto's engine but check the differences.

In 1892 ....
....
....
....

What were the other differences between the two engines?

Practice 4

Read through Study Section 6.1 again. Look for verbs in the past. Make two lists – verbs which use '-ed' (or '-d') and those which do not. Write them next to the present forms and check the spelling.

-ed (-d) past		other	
present	past	present	past
call	called	be	was
use	used	make	made

Study Section 6.2

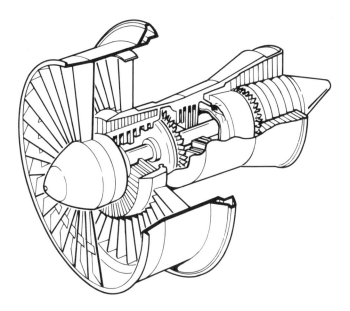

The invention of the internal combustion engine made it possible to fly machines which were heavier than air. The Wright brothers in America began to make and fly gliders in 1900. Then they converted a car engine to drive a propeller and fitted it to a special
5 glider. They first flew their aeroplane on 17th December 1903. They made several flights and the fourth lasted nearly a minute and covered about 250 metres.

For the next forty years aeroplanes were powered by internal combustion engines which drove propellers. In 1941 Frank Whittle,
10 a young English pilot, perfected the jet engine.

The blades of the compressor compressed air which was then passed into a combustion chamber. After fuel injection the burning gases first drove the turbine and were then expelled as a jet. This produced thrust. The turbine drove the compressor which was on
15 the same shaft.

Nowadays, two or more compressors can be used which are driven by two or more turbines for greater power. In turboprop engines the turbine drives a propeller and the jet gives very little forward thrust.

20 Jet engines are very efficient and are now used in all large commercial and military aircraft.

Practice 5

Look at this line from the Study Section:
The Wright brothers in America began to make and fly gliders in 1900. Then they converted a car engine to drive a propeller and fitted it to a special glider.

they refers to the Wright brothers
it refers to propeller
which also refers to other words

What does *which* refer to in the following lines?
Lines 2, 9, 11, 14, 16

Practice 6

Look at these pairs of sentences. Join them correctly and use *which*. The first one has been done for you.

Internal combustion engines made it possible to fly machines.	These drove propellers.
The blades of the compressor compressed the air.	This was then passed to the combustion chamber.
Two or more compressors can be used nowadays.	These are driven by two or more turbines.
For forty years aeroplanes were powered by internal combustion engines.	These were heavier than air.

Example: *Internal combustion engines made it possible to fly machines which were heavier than air.*

Practice 7

Are these statements true or false?
Correct the false ones.

1. The Wright brothers invented the internal combustion engine.
2. The Wright brothers fourth flight lasted 250 minutes.
3. Aircraft had no propellers after 1941.
4. Nowadays all aircraft have jet engines.

Now answer these questions:
1. What made it possible to fly machines heavier than air?
2. What happened in 1900?
   in 1903?
   in 1941?
3. How were the first aircraft powered?

Practice 8

In 1904 the Wright brothers made many flights. Make statements about these flights. Use the notes and the example to help you.

17.01.04./2nd/90 s/755 m
They flew their aeroplane on 17th January, 1904. They made several flights and the second lasted nearly ninety seconds and covered about seven hundred and fifty metres.

1. 18.01.04/3rd/115 s/1100 m
2. 25.02.04/1st/4 min/2.3 km
3. 03.03.04/2nd/5 min 7 s/2.8 km
4. 05.04.04/4th/10 min/4.1 km
5. 09.05.04/1st/12 min/4.8 km
6. 10.05.04/2nd/17 min/5.2 km

Practice 9

Whittle perfected the idea of the jet engine in 1941. Today jet engines use the same principle with more compressors and turbines. These notes describe the construction of a simple jet engine.

Expand the notes into sentences, as in Practice 3, and then put them in the correct order.

1. Burning mixture — pass — the turbine blades — rotate and turn the compressor.
2. A turbine and compressor — connect — a shaft.
3. The blades — compressor compress — air — pass — high speed — the combustion chamber.
4. The shaft rotate — 20,000 rpm.
5. A nozzle spray — fuel — combustion chamber.
6. Burning gases leave — engine — high speed — produce thrust.
7. The fuel — air mixture burn — rapidly — expand.

Now describe Whittle's engine.

In 1941 Whittle perfected the first jet engine. Its design was quite simple. A turbine and compressor were connected to a shaft ....

....
....
....
....

Practice 10

Look at this example:

invention	result
internal combustion engine	fly machines heavier than air

The invention of the internal combustion engine made it possible to fly machines heavier than air.

Now write about these inventions in the same way:

invention	result
television	speak to people over long distances
camera	convert electrical energy into light
radio	record sound
telephone	use steam energy more efficiently
electric light	travel to the moon
steam turbine	produce an electric current
spacecraft	store electrical energy
tape recorder	transmit live pictures
battery	take photographs
generator	transmit sound over long distances

Can you think of other inventions? What were the results?

Study Section 6.3

Space rockets carry a supply of oxygen to travel outside the earth's atmosphere. The fuel can be solid, liquid or gaseous. Many rockets use a liquid fuel, such as kerosene or alcohol, which is pumped with liquid oxygen into the combustion chamber at the rear of the rocket. There the two liquids burn and create a very large volume of hot gases which stream out of the rocket and produce thrust.

A very large force is necessary to lift a rocket off the ground and accelerate it to the speed of about 32,000 km per hour. The exhaust nozzle of the rocket is shaped so that the jet stream moves backwards to produce the large forward thrust on the rocket.

The Saturn 5 rocket which launched the Apollo spacecraft on the journey to the moon weighed more than 3,000 tonnes at take-off. It was a three-stage rocket. The first stage, burning kerosene and liquid oxygen, lifted it 50 km above the earth. This stage then dropped off and the second stage accelerated the two lighter stages to a height of 160 km. The third stage, which, like the second stage, burned hydrogen and oxygen, thrust the spacecraft out of Earth orbit and towards the moon.

Today there are also nuclear rockets. They use a small nuclear reactor to heat a propellant gas such as hydrogen.

Practice 11

Label this diagram with words from Study Section 6.3:

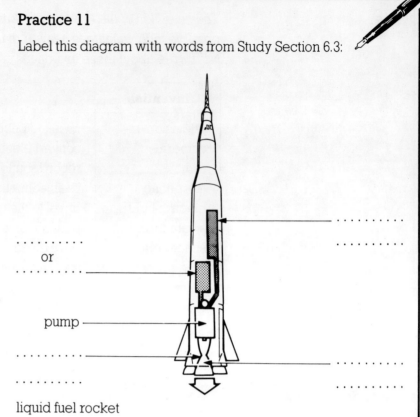

liquid fuel rocket

Practice 12

Are the following statements true or false?
Correct the false ones.
1. Space rockets carry a supply of oxygen because fuels need oxygen to burn.
2. A very large force is necessary to burn the fuel.
3. The rocket moves forwards because the jet stream moves backwards.
4. The Saturn 5 rocket was very heavy because it had three stages.
5. The third stage burned kerosene and oxygen.

Now answer these questions:
1. Why do space rockets carry a supply of oxygen?
2. How is the necessary thrust produced?
3. How high was the spaceship when the first stage dropped off?
4. How do nuclear rockets work?

Practice 13

Look at this sentence:

Many rockets use a liquid fuel such as kerosene or alcohol. *Kerosene* and *alcohol* are examples of liquid fuels.

Now give examples (one or two):
1. The human body has many parts which use the principle of levers such as ....
2. Electric wire is made of metals which are good conductors such ....
3. Aircraft are made of light, tough materials ....
4. Drill bits are made of very tough metal ....
5. Use a metal with a high resistance ....
6. Try a very elastic material ....
7. Use a gas ....
8. Rockets use a liquid fuel ....
9. You need an insulating material ....
10. Short radio waves have a high frequency ....

Practice 14

Make true statements from this table:

A Only a	high small large low	force voltage	is necessary to	cause a light bulb to light. split a nucleus. accelerate a train. open a door. move a finger. power a radio. fly a 747. accelerate a rocket to the moon. move a heavy mass. start a car.

Practice 15

Put in the correct words.

1. The Saturn rocket which .... the Apollo space-craft on the journey to the moon .... more than 3,000 tonnes at take-off.
2. The first stage, ... kerosene and liquid oxygen, ... it 50 km above the earth.
3. For the first forty years aeroplanes were .... by internal combustion engines which .... propellers.
4. The Wright brothers in America .... to make and fly gliders in 1900. Then they .... a car engine to drive a propeller and .... it to a glider.
5. Diesel's first engine .... and nearly killed him but in 1897 he .... successful.

Practice 16

INVENTIONS AND DISCOVERIES

The chart below is mixed up.

Celcius scale	Thompson	1897
simple cell	Hughes	1826
Ohm's law	Bell	1800
electric motor	Benz	1885
electric lamp	Edison	1876
electron	Voltor	1878
telephone	Ohm	1879
motor car	Celcius	1742
microphone	Faraday	1821

Find out the inventor or discoverer and the year.
Make statements:
Benz invented the motor car in 1885.
Ask questions:
Who invented the motor car?

Word list

a carburettor
a design
an exhaust nozzle
an explosion
a flight
a glider
an invention
a jet stream
a nuclear reactor
an orbit
a propellant
a spacecraft
a space rocket
a spark
a supply
a take-off
a tonne
a volume

alcohol
atmosphere
fuel injection
hydrogen
ignition
internal combustion
operation
oxygen
power

efficient
gaseous
liquid
nuclear
solid
turboprop

accelerate
create
design
drop off
expel
explode
fly
ignite
inject
invent
launch
perfect
pump
raise
supply
thrust

Otto designed the first . . .

Unit 7
Electrical Maintenance

Study Section 7.1

How a car battery works

A battery consists of a number of cells. Six-volt batteries have three cells, twelve-volt batteries have six. The cells have positive and negative plates. These are called electrodes and they are immersed in a liquid which is called the electrolyte. Usually the electrolyte is dilute sulphuric acid. Most plates are made of lead.

The battery produces electricity because the acid reacts with the electrodes. Current flows from the negative electrode to the positive electrode through the electrolyte.

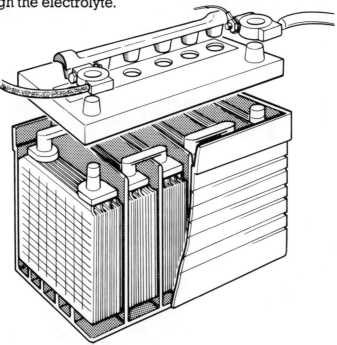

When the reaction between the sulphuric acid and the lead finishes, the electrolyte is very weak. The battery is flat and therefore cannot produce any electricity.

When the battery is recharged with an electric current, the electrodes are converted back to their original condition.

On most cars the earth is connected to the negative terminal, on some to the positive terminal. Check carefully before removing the leads.

Practice 1

Are the statements true or false?
Correct the false ones.
1. A six-volt battery has four cells.
2. The negative and positive plates are the electrodes.
3. Usually the electrode contains water.
4. Most electrodes are made of copper.
5. The reaction of the acid with the electrodes generates the electric current.
6. The current flows from the positive to the negative plate.
7. When the battery is flat the voltage is low.
8. If you recharge a battery the voltage increases.

Practice 2

Answer these questions:
1. What is the voltage of most car batteries?
2. What does a car battery consist of?
3. Why does the battery produce an electric current?
4. When is the electrolyte very weak?
5. Which terminal is the earth connected to on most cars?

Practice 3

Make true statements from this table:

Electricity		sulphuric acid and water.
A battery		iron and carbon.
A jet engine		a blade and a handle.
A hacksaw	consist(s) of	a number of cells.
Ferrous metals		a shaft, a compressor and a turbine.
The electrolyte		a large number of lines.
A television picture		vibrations through the air.
Sound waves		electrons.

Now ask and answer questions with your partner.

For example:
What does a television picture consist of?
It consists of . . . .

Practice 4

Put in the correct form of the verb.

1. The plates (call) electrodes and (immerse) in the electrolyte.
2. The battery (produce) electricity because the acid (react) with the electrodes.
3. When the battery (recharge) with an electric current, the electrodes (convert) back to their original condition.
4. On most cars the earth (connect) to the negative terminal.

Language Point

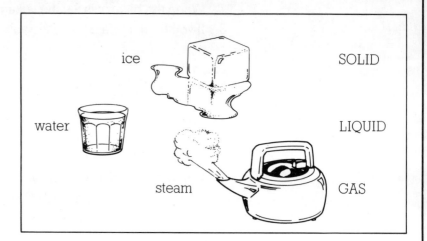

Practice 5

Put the substances in the correct column.
Add other substances.

solid	liquid	gas
	water	

hydrogen, acid, rubber, air, petrol, coal, oil, steam, copper, aluminium, carbon, carbon dioxide, sulphuric acid, electrolyte, distilled water, . . .

Study Section 7.2

Care and maintenance of the battery

A battery should be kept dry and clean. It should be fixed firmly to prevent the electrolyte from spilling out.

If the electrolyte level is too high it can spill out and the terminals can become corroded with acid. Current can discharge across dirty and damp surfaces. Spilled electrolyte can damage the brackets which secure the battery.

Materials

Grease
Emery paper
A rag
Distilled water

Tools

A screwdriver
A spanner

Disconnect the terminals with a spanner. If the terminals are tight, tap with a screwdriver handle.

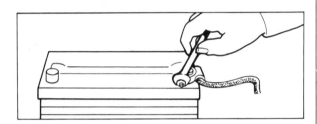

Clean the terminals with a piece of emery paper and dry with a clean rag.

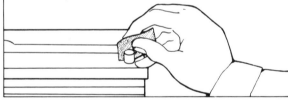

Smear the terminals with grease to prevent corrosion. Replace and tighten the nuts.

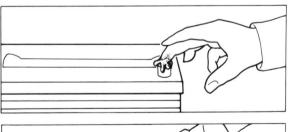

Top up the battery with distilled water until the electrolyte covers the plates.

Practice 6

Are these statements true or false?
Correct the false ones.

1. A battery should be kept dry and clean.
2. It should be fixed to prevent the acid spilling out and corroding the terminals.
3. Current can leak across clean surfaces.
4. Spilled electrolyte causes metals to corrode.
5. Battery terminals should be cleaned with emery paper.
6. Batteries should be topped up with distilled water until the acid covers the electrodes.

Now answer this question:
Why should a battery and its terminals be kept dry and clean?

Language Point

Keep the battery clean!
　　The battery should be kept clean.

Fix the battery firmly!
　　It should be fixed firmly.

keep	be kept
top up	be topped up
dry	be dried

Practice 7

Make statements about the pictures.
Use *should*.

Look at this example:
The battery should be kept clean.

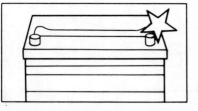

keep

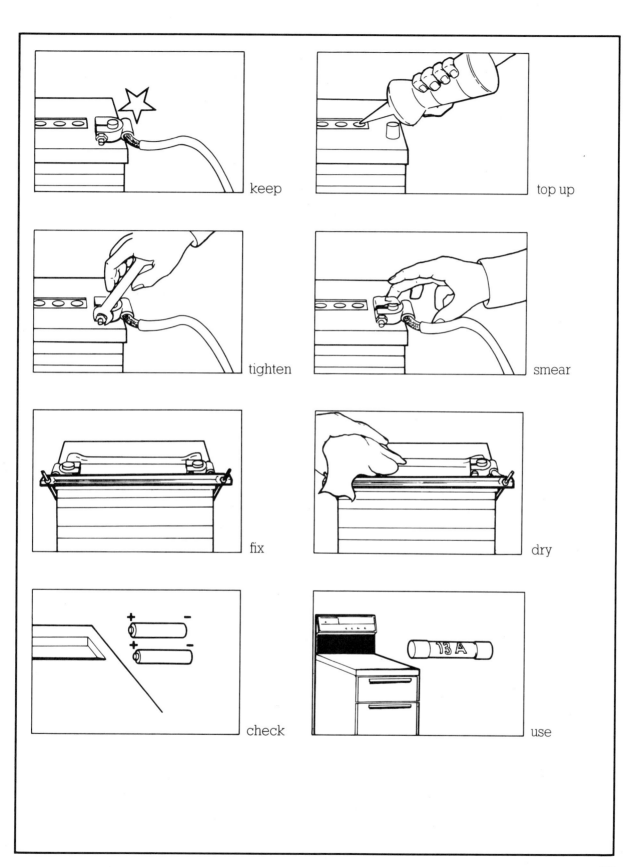

Now make statements about servicing a car.

For example:
The battery should be topped up.
The tyre pressures should be checked.
(the brakes, the oil level, the lighting system, the steering, the spark plugs, the fan belt, etc.)

Practice 8

Give instructions.

For example:
If the terminals are tight, *tap with a screwdriver handle.*

1. If the terminals are dirty, ...
2. If the bolt is loose, ...
3. If the fuse rating is too low, ...
4. If the car will not start, ...
5. If the drill will not cut, ...
6. If the brake fluid level is too low, ...
7. If the radio is too quiet, ...
8. If the lamp is too dim, ...
9. If the tyre is too worn, ...
10. If the battery is flat, ...

Study Section 7.3

Checking the battery

The condition of the battery depends on the strength of the electrolyte. It becomes weaker when the battery discharges and stronger when being charged by the car generator.

Check the battery condition by taking some electrolyte from the cells with a hydrometer.

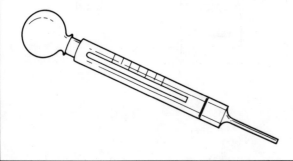

The hydrometer measures the strength of the acid in the cells. This is called the specific gravity. It should be 1.260. If the hydrometer reading is 1.260 to 1.280, the battery is fully charged. If the reading is below 1.100, the battery is flat.

Squeeze the bulb on the hydrometer. Insert the nozzle into a cell. Release the bulb to draw electrolyte into the hydrometer.

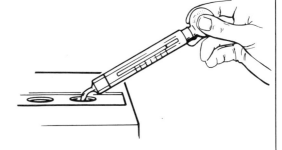

Check the reading. Return the fluid to the cell and check the other cells.

Practice 9

Are these statements true or false?
Correct the false ones.

1. If the electrolyte is weak the battery is fully charged.
2. The electrolyte becomes stronger when the battery discharges.
3. A hydrometer is used for measuring specific gravity.
4. If the hydrometer reading is 1.000 the battery is without charge.
5. A battery is flat when the specific gravity of the electrolyte is too low.

Now answer the questions:

1. What does a hydrometer do?
2. What does the strength of the battery depend on?
3. What is the hydrometer reading for a fully-charged battery?
4. When is the battery flat?

Practice 10

Make true statements about the condition of the battery.
For example:
1.260
If the hydrometer reading is 1.260, the battery is fully charged.
1 1.270   2 1.090   3 1.280   4 1.080   5 1.090

Language Point

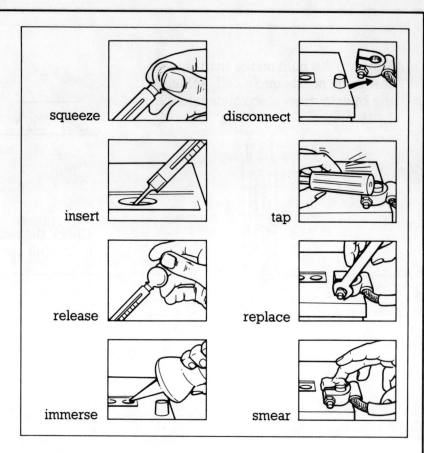

Practice 11

Put in the correct instruction words.

1 .... the terminals with grease.
2 .... the rubber bulb of the hydrometer.
3 .... the lead from the terminal.
4 .... the nut if it is too tight.
5 .... the electrodes in the electrolyte.
6 .... the nozzle into the holes.
7 .... and tighten the nuts.
8 .... the bulb to draw up the electrolyte.

Now make sentences with *should*.
Loosen the nut before disconnecting the wire.
The nut should be loosened before disconnecting the wire.

Practice 12

Put in the right words.
Use:
of, with, from, on, into, across, to, through
You can use some words twice.

1. A battery consists .... a number .... cells.
2. The acid reacts .... the electrodes.
3. Current flows .... the negative electrode .... the positive .... the electrolyte.
4. The electrodes are converted back .... their original condition.
5. The terminals will become corroded .... acid.
6. Current can discharge .... dirty and damp surfaces.
7. Disconnect the terminals .... a spanner.
8. The strength .... the battery depends .... the strength .... the electrolyte.

Practice 13

Put in the correct words.

A car battery .... of a number of cells which have .... and negative plates. These are called .... and are immersed in .... sulphuric acid. The battery .... current because the acid .... with the electrodes. When the electrolyte is very .... the battery has no power. However, when the battery is .... the electrodes are .... back to their original condition.

To keep the battery in good condition it .... be kept dry and clean. If it is not .... firmly, electrolyte can .... out and the .... can become .... with acid. Spilled electrolyte can also .... the brackets which secure the battery.

To prevent current from .... across dirty terminals they should be .... with emery paper and .... with grease to prevent corrosion. Always .... up the battery with .... water to maintain the battery in good condition.

Word List

a bracket	firmly
a bulb	
a cell	charge
an earth	check
an electrode	consist of
a hydrometer	corrode
a lead	damage
a liquid	dilute
a plate	discharge
a reaction	disconnect
a reading	fix
a surface	immerse
a terminal	keep
	prevent
corrosion	recharge
distilled water	replace
electrolyte	secure
grease	service (a car)
specific gravity	smear
strength	spill out
sulphuric acid	tap
	tighten
damp	top up
dirty	
flat	should be fixed/kept etc
level	
negative	
positive	

Unit 8
Technical Books and Reports

Study Section 8.1

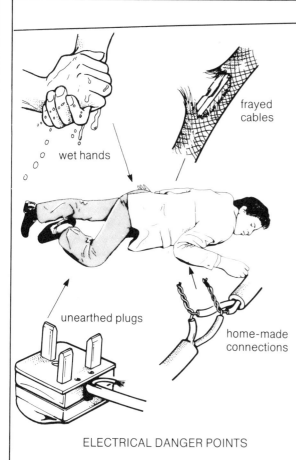

DON'T use defective electrical equipment.

REPORT faulty electrical equipment immediately.

NEVER touch electrical equipment with wet hands.

REPORT all frayed cables.

DON'T make faulty electrical connections.

USE the advice of an authorised electrician.

CHECK the earthing on all power tools.

ELECTRICAL DANGER POINTS

Practice 1

Are these statements true or false?
Correct the false ones.

1. Defective electrical equipment is safe.
2. Frayed cables are well insulated.
3. Power tools should be earthed.
4. Wet hands conduct electricity easily.
5. An electrician should check faulty electrical devices.

Practice 2

Now change warnings into instructions. Use *must*.
Look at the example:

| Don't use | defective electrical equipment. |

Defective electrical equipment | must | not | be | used.

Then make true statements from the table.

Faulty electrical equipment			called.
An authorised electrician			reported immediately.
Unearthed plugs			replaced.
Electrical equipment			used.
Faulty brakes			tightened.
Faulty electrical connections			recharged.
Frayed cables	must	(not)	touched with wet hands.
Damaged tools		(never) be	checked regularly.
A flat battery			measured accurately.
A loose nut			repaired.
A broken fuse			adjusted.
A defective machine			made.
A broken light bulb			worn.
Voltage			
Correct clothing			
A faulty switch			
Worn tyres			

Ask and answer questions about the table above.
Look at the example:
Why were the cables replaced?
Because they were faulty.

Practice 3

Look at this:

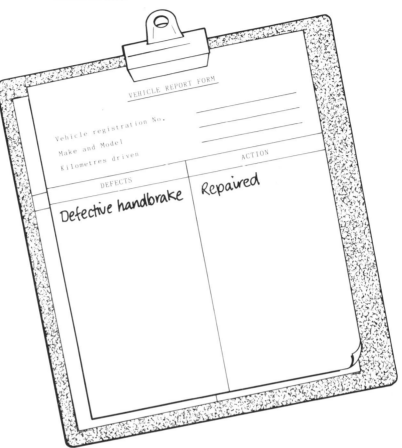

Now look at this car and enter its details in the report form above. Look at the faulty parts. Must the defective or faulty parts or systems be repaired, adjusted or replaced? Put this information in the report form.

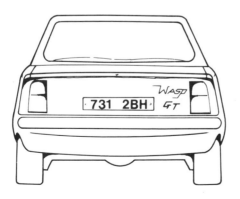

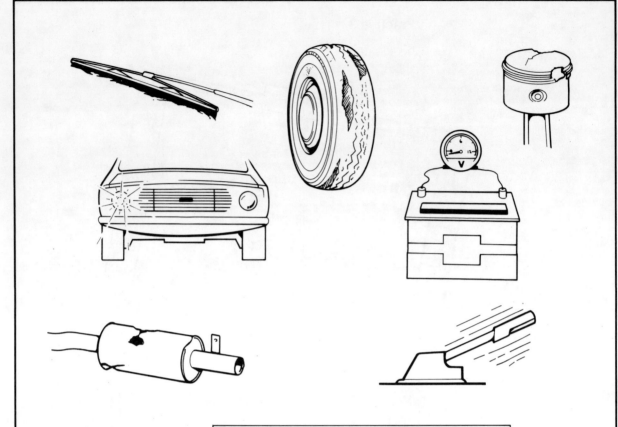

Language Point

≃ The handbrake must be repaired.
The handbrake needs repairing.

Practice 4

Now look at the completed report form and make true statements about the car.

Look at the examples:
The defective handbrake needs repairing.
The handbrake is defective and must be repaired.

Practice 5

Look at these descriptions of problems:
The electrolyte level is too low and needs topping up.
Now use the notes to make similar statements.

1. The fuel level .... low .... top up.
2. The temperature .... high .... lower.
3. The engine .... hot .... cool down.
4. The voltage .... high .... decrease.
5. The drill speed .... low .... increase.
6. The volume .... loud .... turn down.
7. The tyres .... worn ....
8. The fuse rating .... high ....
9. The amount of exhaust ....
10. The switch .... high ....

Now read through this page from an Introductory Handbook for Craft Technicians.

Study Section 8.2

HIGH CARBON STEEL

high carbon steel

showers of sparks

High carbon steel is hard and it can be recognised in the spark test by a shower of sparks produced by the high carbon content. It is used for engineering tools such as chisels and drills.

ALLOYS OF STEEL

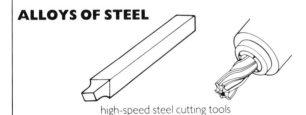

high-speed steel cutting tools

The properties of steel can be improved by the addition of other elements such as tungsten. Alloy steels have great strength and toughness and can be used at high temperatures. They are used for cutting tools.

COPPER

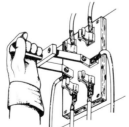

heavy switch gear

Pure copper is very soft and can be made into sheet, wire and tubes. It conducts electricity very easily and can be used for electrical wire; it is also used for hot water tanks and modern plumbing.

Practice 6

Are these statements true or false?
Correct the false ones.

1. When a chisel made of high carbon steel is sharpened, a shower of sparks is produced.
2. Drill bits are usually made of copper because of its hardness and toughness.
3. Tool steel is an alloy of steel and tungsten.
4. Alloy steels have a high temperature.
5. Pure copper is usually quite hard.
6. Copper is used for modern plumbing because it is a good conductor of electricity and heat.

Now try this:
Compare high carbon steel and mild steel. What is mild steel used for and why? (Look at the information on page 89.)

Now talk about the properties of other materials. What are they used for and why?

Practice 7

Are these statements true or false?
Correct the false ones.

1. Tool steel can be scratched very easily.
2. Rubber can be stretched very easily.
3. Gold can be recognised by its colour.
4. The properties of diamonds can be improved by the addition of other elements.
5. Worn tyres can be improved by painting.
6. FM or very high frequency radio waves can be transmitted over long distances.
7. Worn clutches can never be repaired. They must be replaced.
8. Steel can be made harder by heating and rapid cooling.
9. The air temperature inside a piston can be raised by greater compression.
10. The condition of a battery can be improved by cleaning and maintenance.

Study Section 8.3

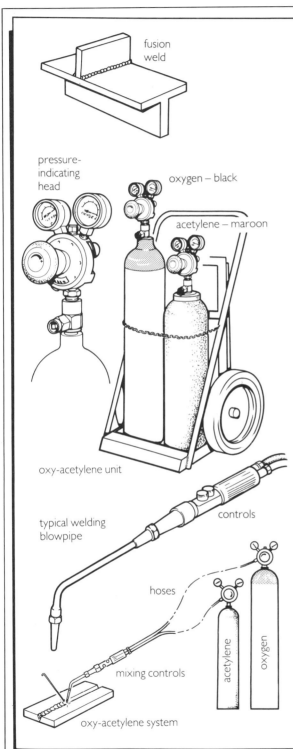

WELDING

Welding is the joining of metals by fusion.

METHODS OF WELDING

Oxy-acetylene

When oxygen and acetylene gases are mixed correctly they give a flame which reaches a temperature of approximately 3,300 °C.

In a high-pressure system, the oxygen and acetylene are used from high-pressure cylinders. The acetylene should be adjusted to working pressure first and then the oxygen. In a low-pressure system the acetylene is generated in a low-pressure generator.

The oxygen and acetylene are carried by hose to the blowpipe. Nozzles with different sizes are used to change the heat of the flame. These are marked to show the gas consumption. It is important to use a nozzle of the correct size, depending on the thickness of the metal. A small flame from a large nozzle must never be used. The flame must be adjusted correctly for good welds. Too much oxygen causes brittleness.

Practice 8

Make correct statements from the table.

The nozzles are marked	must be adjusted correctly.
It is important	are used to adjust the heat.
Welding is	causes brittleness.
A small flame from a large nozzle	the acetylene is generated in a low pressure cylinder.
Too much oxygen	must never be used.
Nozzles of different sizes	the joining of metals by fusion.
In a low pressure system	to show gas consumption.
The flame	to use a nozzle of the correct size.

Practice 9

1. What is welding?
2. What is the temperature of a correct mixture of oxygen and acetylene?
3. Which gas should be adjusted for pressure first?
4. How is the heat of the flame adjusted?
5. What does the correct nozzle size depend on?
6. How is brittleness in the weld caused?
7. What do the gauges on the cylinders show?
8. How is the acetylene generated in a low pressure system?

Study Section 8.4

PROTECTIVE DEVICES

The cables which carry electric current to the different appliances in the factory are called conductors. Their resistance to the flow of electric current causes heat to be generated. If the flow of electricity in a circuit suddenly increases, the heat in the conducting wires increases and can cause the insulation to burn, resulting in damage to the equipment and possibly causing a fire. All electric circuits must therefore be protected by manual trip switches, fuses or automatic trip switches, which cut the supply where there is a fault.

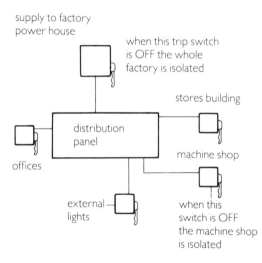

MANUAL TRIP SWITCHES

Manual trip switches can be inserted in the circuit and are used to isolate equipment from the supply if a fault occurs.

FUSES

Every circuit should have a fuse in the supply line to the equipment. The fuse can be of the cartridge type which is held in place by two spring clips or it can be a piece of wire joining two points in the circuit. Fuses are rated in amperes and the rating must be correct so that if there is an overload the fuse will melt and break the circuit before any damage to the circuit wiring or the equipment occurs. If a fuse blows, the correct manual switch must be switched off and the fault rectified before replacing the fuse. When a fuse is changed, the correct rating must be used. It is dangerous and can be expensive to fit a replacement fuse with a higher rating.

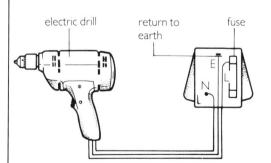

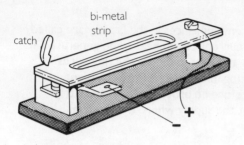

AUTOMATIC TRIP SWITCHES

Automatic trip switches are often fitted to electrical devices to protect them from overloads. Usually, they are thermal switches which break the circuit when an overload causes a bi-metal strip in the switch to bend. These switches have a re-set button which re-sets the trip switch and completes the circuit. If the circuit is broken because of an increase in current, the switch cannot be re-set immediately. The bi-metal strip needs time to cool (approx. 30 seconds) before the switch can be re-set.

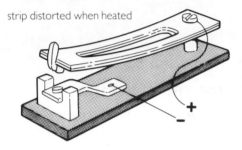

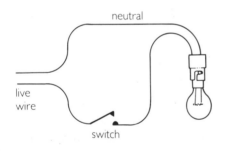

CONNECTING PLUGS

A lamp is connected with the switch open. The lampholder can only become live when a fault occurs in the insulation which is not isolated by the switch.

When plugs for operating electrically-powered hand tools are connected, the maximum rating for the connector must not be exceeded.

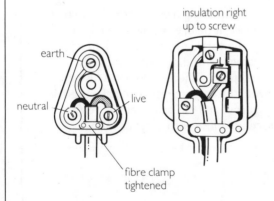

Practice 10

Are these statements true or false?
Correct the false ones.

1. Electric circuits must be protected against current overloads.
2. An increase in current causes heat to be generated in conductors.
3. Burning insulation can be caused by fuses.
4. Defective circuits must be isolated from the supply.
5. Fuses can be replaced immediately after melting. Use a higher rating if necessary.
6. A bi-metal strip consists of one piece of metal.
7. A bi-metal strip must cool before the circuit is complete again.
8. The correct manual switch must not be switched off before replacing the fuse.

Practice 11

Put in the correct words.
Choose from this list:

rectified	piece	must	blows	held
damage	should	melts	manual	cartridge
joining	clips	breaks	occurs	
supply	rated	of	there	

Every circuit .... have a fuse in the .... to the equipment. The fuse can be of the .... type, which is .... in position by two spring ...., or it can be a .... of wire .... two points in the circuit. Fuses are .... in amperes and the rating .... be correct so that if .... is an overload .... current, the fuse wire .... and .... the circuit before any .... to the circuit wiring .... If a fuse .... the correct .... switch must be switched off and the fault .... before replacing the fuse.

Practice 12

Explain the difference between automatic trip switches, manual switches and fuses and why all electrical circuits must be protected against current overloads.

Examine the electrical systems and appliances in your school or college. Draw diagrams of the circuits and supply. Describe all the protective devices which you find.

Word List

a blowpipe
a cable
a cartridge
a connection
a hose
an overload
a trip switch

acetylene
alloy steel
earthing
equipment
fusion
high pressure
low pressure
plumbing
welding

automatic
bi-metal
defective
faulty
frayed
manual
protective
pure
thermal

cause something to . . .
insert
isolate
occur
protect
rectify
result in

Don't . . .
It must be repaired/fixed . . .
It needs repairing/fixing . . .

Word List
Elementary Technical English 2

The number after each word indicates the unit where the word is first used. For example:

accelerate (v) 6

— means that the word **accelerate** is first used in Unit 6. Most of the words in this list are introduced in this book and do not appear in Book 1.

The letters in brackets ( ) indicate if the word is:
- an adjective (adj)
- an adverb (adv)
- a conjunction (conj)
- a preposition (prep)
- a verb (v)
- a countable noun (nc)
- an uncountable noun (nø)
- a plural noun (npl)

A countable noun is a noun which we can count. For example:

an aerial (nc)   one aerial, two aerials, six aerials.

An uncountable noun is a noun which we can't count. For example:

acetylene(no)   acetylene, some acetylene

Therefore:

accelerate (v) 6 – is a verb and it is first used in Unit 6.
acetylene (no) 8 – is an uncountable noun and it is first used in Unit 8.

accelerate (v) 6
acetylene (nø) 8
adjust (v) 4
aerial (nc) 5
aileron (nc) 1
aircraft (nc) 1
alcohol (nc + ø) 6
alloy steel (ø) 8
alternate (v) 5
alternating current (nc + ø) 5
alternating cycle (nc) 5
ammeter (nc) 4
amp (nc) 4
amplifier (nc) 5
amplify (v) 5
anti-clockwise (adj) 2
apply (v) 2
approximately (adv) 5
atmosphere (nc + ø) 6
automatic (adj) 8

backwards (adv) 5
balanced (adj) 2
battery (nc) 4
because (conj) 3
bi-metal (adj) 8
blade (nc) 1
blowpipe (nc) 8
boiler tube (nc) 4
bolt (nc) 3
bracket (nc) 7
bright (adj) 5
brittle (adj) 3
brittleness (nø) 3
bulb (nc) 7
bunker (nc) 4
burn (v) 1

cable (nc) 8
calculate (v) 4
carbon (nø) 3
carburettor (nc) 6
cartridge (nc) 8
cast iron (nø) 3
cause something to . . . (v) 8
cell (nc) 7
change shape (v) 3
charge (v) 7
check (v) 7
circuit (nc) 4
clockwise (adj) 2
coil (nc) 5
combine (v) 5
combustion chamber (nc) 1
commercial (adj) 1
compress (v) 1
compression (nc) 1
conductor (nc) 3
connection (nc) 8
consist of (v) 7

construction (nc + nø) 1
construction (nc) 1
contain (v) 3
control surface (nc) 1
convert (v) 5
conveyor belt (nc) 4
cool (v) 4
copper (nø) 3
corrode (v) 7
corrosion (nø) 7
crane hook (nc) 3
create (v) 6
cutting tool (nc) 3
cycles per second (npl) 5
cylinder (nc) 1

damage (v) 7
damp (adj) 7
defective (adj) 7
design (nc) 6
design (v) 6
device (nc) 4
diaphragm (nc) 5
dilute (v) 7
dim (adj + v) 4
direct current (nc + ø) 5
dirty (adj) 7
discharge (v) 7
disconnect (v) 7
distilled water (nø) 7
divided by (prep) 2
downwards (adv) 5
drop off (v) 6

earth (nc) 7
earthing (nø) 8
efficent (adj) 6
effort force (nc) 2
elastic (adj) 3
elasticity (nø) 3
electric current (nc + ø) 4
electrical (adj) 1
elecetrical energy (nc + ø) 4
electrode (nc) 4
electrolyte (nc + ø) 7
electron (nc) 5
elevator (nc) 1
emery paper (nø) 4
engineering material (nc) 3
equal (adj + v) 2
equipment (nø) 8
eraser (nc) 3
evert (v) 2
exhaust nozzle (nc) 6
expand (v) 1
expel (v) 6
explode (v) 6
explosion (nc) 6

faulty (adj) 8

ferrous (adj) 3
ferrous metal (nc + ø) 3
firmly (adv) 7
fix (v) 7
flat (adj) 7
flight (nc) 6
flow (nc + ø) 4
fly (v) 6
force (nc) 2
formula (nc) 4
forwards (adv) 5
frayed (adj) 8
frequency (nc) 5
fuel (nc + ø) 1
fuel injection (nø) 6
furnace (nc) 4
fuse (nc) 5
fusion (nø) 8

gap (nc) 4
gaseous (adj) 6
generate (v) 4
generator (nc) 4
girder (nc) 3
glider (nc) 6
grease (nc + ø) 7
grind (v) 4
guitar (nc) 5

hard (adj) 3
heat (v) 4
heating element (nc) 4
heavy (adj) 3
helicopter (nc) 1
hertz (nc + ø) 5
high carbon steel (nø) 3
high pressure (nø) 8
hose (nc) 8
however (conj) 3
hydraulic (adj) 1
hydrogen (nø) 6
hydrometer (nc) 7

ignite (v) 6
ignition (nø) 6
immerse (v) 7
inject (v) 6
insert (v) 8
insulation (nø) 4
insulator (nc) 3
internal combustion (nø) 6
invent (v) 6
invention (nc) 6
iron (nø) 3
isolate (v) 8

jet engine (nc) 1
jet stream (nc) 6

keep (v) 7
kerosene (nc + ø) 1
kilohertz (nc + ø) 5

99

lateral (adj) 1
launch (v) 6
lead (nc) 7
level (adj) 7
lever (nc) 2
lift (nc + ø) 1
light (adj) 3
light wave (nc) 5
liquid (adj + nc) 6
load (nc) 2
longitudinal (adj) 1
loudspeaker (nc) 5
low pressure (nø) 8

magnify (v) 2
mains supply (nc) 4
manual (adj) 8
measure (v) 4
mechanical (adj) 1
mechanical advantage (nc) 2
medium carbon steel (nø) 3
megahertz (nc + ø) 5
melt (v) 5
metal (nc + ø) 3
microphone (nc) 5
mild (adj) 3
mild steel (nø) 3
military (adj) 1
minus (prep) 2
moment (nc) 2

negative (adj) 7
Newton metre (nc) 2
nichrome (nø) 4
nozzle (nc) 1
nuclear (adj) 6
nuclear reactor (nc) 6
nut (nc) 3

occur (v) 8
Ohm (nc) 4
operation (nc + ø) 6
orbit (nc) 6
overload (nc) 8
overtighten (v) 4
oxygen (nø) 6

pass through (v) 4
perfect (v) 6
piston engine (nc) 1
pivot (nc) 2
plate (nc) 7
plumbing (nø) 8

plus (prep) 2
positive (adj) 7
power (nø) 6
power rating (nc) 4
prevent (v) 7
produce (v) 1
propellant (nc) 6
propellor (nc) 1
property (nc) 3
protect (v) 8
protective (adj) 8
pump (v) 6
pure (adj) 8

quantity (nc + ø) 3

radiate (v) 5
raise (v) 6
rapidly (adj) 1
reaction (nc) 7
reading (nc) 7
receive (v) 5
receiving aerial (nc) 5
recharge (v) 7
rectify (v) 8
regain (v) 3
replace (v) 7
resist (v) 4
resistance (nø) 4
resistor (nc) 4
result (nc) 5
result in (v) 8
ripple (nc) 5
rotate (v) 1
rotational (adj) 1
rotor blade (nc) 1
rubber (nc) 3
rudder (nc) 1

scratch (v) 3
screen (nc) 5
secure (v) 7
see-saw (nc) 2
service (a car) (v) 7
shaft (nc) 1
smear (v) 7
soft (adj) 3
solid (adj) 6
sound wave (nc) 5
space rocket (nc) 6
spacecraft (nc) 6
spark (nc) 6

specific gravity (nø) 7
speed of light (nø) 5
spill out (v) 7
spray (v) 1
steel (nc + ø) 3
strength (nø) 7
strong (adj) 3
studio (nc) 5
sulphuric acid (nø) 7
supply (nc + v) 6
surface (nc) 7
switch (nc) 4

take-off (nc) 6
tap (v) 7
terminal (nc) 7
therefore (conj) 3
thermal (adj) 8
thrust (nø + v) 1
tighten (v) 7
times (prep) 2
tonne (nc) 6
tool steel (nø) 3
top up (v) 7
tough (adj) 3
transmit (v) 5
transmitter (nc) 5
transmitting aerial (nc) 5
tripswitch (nc) 8
tungsten (nø) 3
tuning knob (nc) 5
turbine (nc) 1
turboprop (adj) 6
turning force (nc) 2

up to . . . (prep) 1
upwards (adv) 5

vibrate (v) 5
vibration (nc) 5
volt (nc) 4
voltage (nø) 4
voltmeter (nc) 4
volume (nc) 6

watt (nc) 4
wave (nc) 5
wavelength (nc) 5
weak (adj) 3
welding (nø) 8
which (conj) 3
wrought iron (nø) 3